WELCOME TO NEUROVOY INDUSTRIES

Welcome to NeuroVoy Industries

New Employee Handbook

DEREK A. MCKAY

CONTENTS

Dedication   vii

1. Welcome and Introduction   1
2. Offices, Primary Employees, Support Staff, and Roles   7
3. Our Building   19
4. Major Departments and Teams   33
5. Important Committees   55
6. Potential Complications   64
7. Good Luck!   86

Human Glossary and Index   87
Author's Note and References   109
Acknowledgments   111
About the Author   113
For More Information   115

Copyright © 2024 by Derek A. McKay

All rights reserved. No part of this book may be reproduced in any manner whatsoever without written permission except in the case of brief quotations embodied in critical articles and reviews.

This is dedicated to my son, Colin Voy, and my wife, Christine. You are my guiding lights. I love you both.

| 1 |

Welcome and Introduction

Hello esteemed nucleus! Our Department of Employee Resources (DER) would like to welcome you to NeuroVoy Industries. We are excited for you to join our staff. It is not every day that we are able to hire a new nucleus such as yourself. You probably have many questions and we hope this handbook will help to answer them. The primary purpose of this handbook is to introduce and orient you to:

1. The offices, roles, and functions of employees within our company.
2. The layout of our building.
3. The major departments and teams within our company.
4. Several important committees within our company.
5. Potential complications that you, our company, our clients, and even our building may encounter.

Our primary clients are called "humans," but we also provide services for other creatures as well. We will describe things in this handbook using both NeuroVoy terminology and some primary terms that humans use (color-coded in blue). Ultimately, NeuroVoy Industries works for a parent company known as NS Enterprises. To humans, this is known, broadly, as the "nervous system." NS Enterprises has two companies working for them. One company is ours and the other is a company called Periphalink Solutions. You can see the organizational structure in Figure 1.1 below.

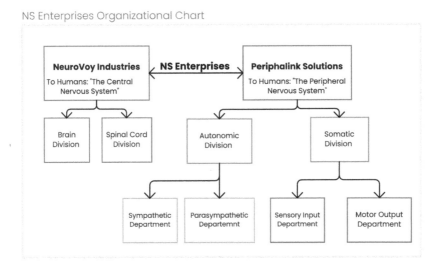

Figure 1.1 - Organizational Structure of NS Enterprises

To humans, NeuroVoy Industries is known as the "central nervous system." We are divided into the Brain Division (brain) and the Spinal Cord Division (spinal cord) and so we are responsible for the function and monitoring of these areas for our clients. You have been hired to work in the Brain Division. Meanwhile, Periphalink Solutions deals with everything outside the brain and spinal cord. In other words, they deal with things in the "periphery" and humans call this company the "peripheral nervous system." It is important to note that these two companies work very closely and very well with each other.

Periphalink Solutions is broken down into an Autonomic Division (autonomic nervous system) and a Somatic Division (somatic nervous system). The Autonomic Division deals with things like regulating a human's internal organs and other bodily functions. Digestion is a great example of this. When you think Autonomic, think "automatic." Meaning, digestion happens, and humans do not have to think consciously about digesting their lunch. The Autonomic Division has two departments known as the Sympathetic Department (sympathetic nervous system) and Parasympathetic Department (parasympathetic nervous system). When humans feel threatened or are experiencing high levels of stress, the Sympathetic Department may take the lead to send resources out to their body and brain to take on this challenge. For example, if a human walked outside of their house and saw two hungry lions sitting there (Figure 1.2), the Sympathetic Department will push resources towards areas of the body and brain that would help the human either fight or flee from the lions. Some humans know this

Figure 1.2 - Two Hungry Lions

function as the "fight or flight response." The Sympathetic Department would help the human to increase their heart rate, release hormones, and decrease digestion. Once a threat or stressful event is over or reduced, the Parasympathetic Department may come in to finish the job. This department helps to calm the body and brain back to a nice, even level. So, let's say a human is successfully able to run back inside their house and away from the lions. The Parasympathetic Department would help humans to decrease their heart rate, stop production of certain hormones, and let the human go back to digesting their lunch. Again, it is important to note that these departments are working automatically and without a human's conscious direction. Humans do not consciously will their body to produce certain hormones, for example.

The other major division within Periphalink Solutions is the Somatic Division, which allows humans to have more conscious control over some functions. When you think of the term "somatic," think "body." This division has two departments. The Sensory Input Department takes sensory information and inputs it into the human body. This includes sounds, smells, taste, and touch. This information is then sent to NeuroVoy Industries for processing. As a side note, sight is processed a bit differently and does not need to connect through the Sensory Input Department. The Motor Output Department receives impulses from NeuroVoy and helps to control voluntary muscle movements. For example, it allows humans to walk, run, swim, grab a fork to eat, type on a computer, etc. The term "motor" relates to "movement." This is a basic overview of NS Enterprises but be sure to reach out to your supervisor or DER if you would like more information. Figure 1.3 is a promotional poster that was designed for humans awhile back to show a basic overview and the processes of NS Enterprises.

You may be wondering how you came to be, how old you are, and more. Well, it depends on the client. For example, our client is 36 years old and so many of our employees are 36 years old. So is our building and many of our committees, too. Other clients might only be 5 years old or even 77 years old. So, it really just depends. But occasionally, we are able to hire new employees! This is where you come in. You likely came about by a process called "neurogenesis." We do not necessarily have full control over your creation. Our clients can actually impact whether you exist or not. For example, some of our clients consistently engage in things like physical exercise, cognitively stimulating activities, and get good, restorative sleep. So, it may be the case that you exist and are being hired by NeuroVoy, in part, because of our client. Of

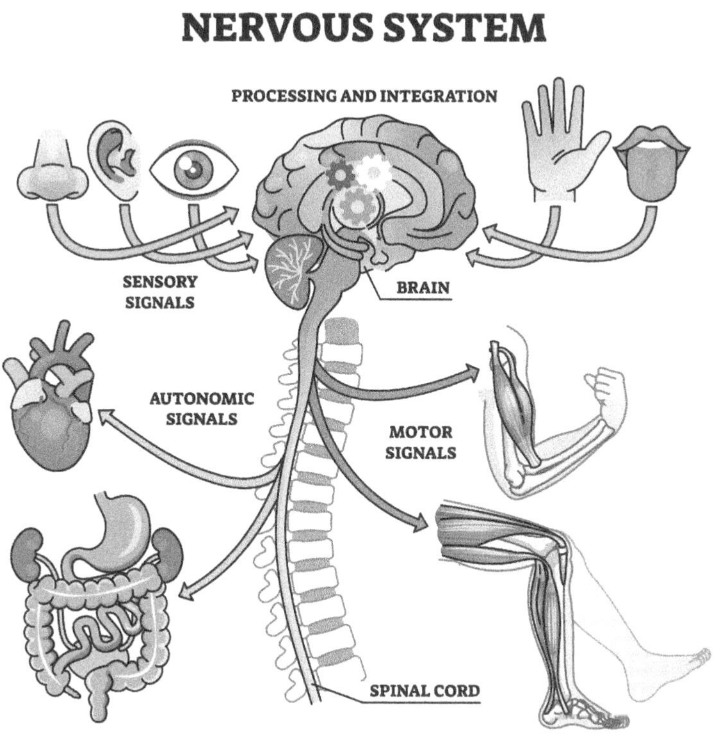

NERVOUS SYSTEM

Figure 1.3 - NS Enterprises Promotional Poster

course, there is likely more to it, but there is always new research being done to determine the process of neurogenesis!

Before we move to other portions of the handbook, we would like to discuss what humans term "cognition." Cognition, broadly, refers to the way that our clients think, acquire knowledge, remember, and so on. We will reference several domains of cognition throughout the handbook, and we have provided the way that humans think about and have described these different areas as follows:

- Sensorimotor functioning generally describes the ability to use sensory information and integrate it into motor function.
- Attention generally refers to the ability to receive and process incoming information.

- Processing speed generally refers to the ability to use that information quickly and efficiently.
- Language functioning generally refers to the ability to express and receive communication.
- Visuospatial functioning generally refers to the ability to identify and process visual information and spatial relationships.
- Memory generally refers to the ability to input, store, and retrieve information.
- Executive functioning generally refers to higher order cognition including ability to plan, make judgements, reason, make decisions, solve problems, etc.

Further, as you make your way through the handbook, you will notice the human term "disorder" used. We here at NeuroVoy Industries do not prefer the term "disorder" due to the inherent stigma attached to this term. We prefer to use the term "diagnosis" or "diagnoses." However, when referring to human lingo and to keep things consistent, you will still see the term "disorder" occasionally used.

We are extremely excited for your arrival and are happy to have you on the team! As you settle into your role, know that you are surrounded by a supportive and dedicated team ready to assist you in any way possible. Once again, welcome to NeuroVoy Industries. We look forward to building great things together, achieving milestones, and creating lasting memories as we embark on this journey as a cohesive and thriving team. As you progress through your career here, we will likely have more handbooks for you to review so please stay tuned!

| 2 |

Offices, Primary Employees, Support Staff, and Roles

You, like many of our primary employees, will be assigned to a specific office. To humans, this office is known as a "neuron." There are roughly 100 billion offices working for one client and our employees all play a vital role in our success. These offices, though, vary in terms of how they look (anatomy) and what work is being done in them (function). This is an important point because these differences allow our employees to perform a wide array of tasks. Sometimes, offices can even be altered depending on the needs of our clients which creates another layer of complexity in terms of what our employees can do. In this section of the handbook, we will describe the main components and general processes of an office, support staff you may encounter, your office, and your role, specifically.

The Main Components/General Processes of an Office

Figure 2.1 is an image of an office (a multipolar neuron) with one of our employees (the nucleus) named Kevin (with his consent, of course).

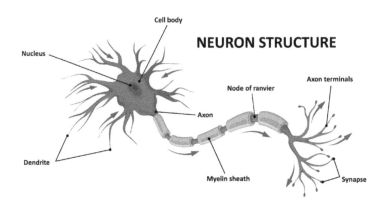

Figure 2.1 - Kevin's Office

Before we get to Kevin, there are a few key points and terms we would like to identify for you. The first point is that many of our employees send information to each other. This information comes in the form of printed memos. The information contained in our memos help the receiving employee figure out what they need to do next, and it is a primary method of communication. Humans call these memos "neurotransmitters," and they are chemical in nature. Some common neurotransmitters include dopamine, serotonin, acetylcholine, GABA, glutamate, oxytocin, and norepinephrine. Neurotransmitters are involved in so many functions! Broadly, they help influence how humans learn and remember things, regulate mood, move their bodies, respond to stress, interact with others, and help keep internal processes balanced. Figure 2.2 shows one of our posters for a neurotransmitter promotion we ran last year for our client. His child was recently born, he got a new job, and we were trying to promote exercise and good sleep! Our employees strive to send these memos to each other in the proper fashion, but problems can happen. For example, humans sometimes have medical conditions that can disrupt the communication of information (more on that later). Other things like drugs or medications can also influence how neurotransmitters act, either by increasing an action (humans call these "agonists") or decreasing an action (humans call these "antagonists").

OK, back to Kevin. In the left of Figure 2.1, you will notice arrows coming towards the neuron (the office). Those arrows signify printed memos (neurotransmitters) being passed into Kevin's office from another employee. In this case, the other employee is one of Kevin's colleagues named Stephanie. Stephanie has sent some information about their client's leg movement. Memos have to be dropped off somewhere and so Kevin has a few mailboxes outside of his office to collect them.

Figure 2.2 - Neurotransmitter Promotional Poster

These mailboxes are known to humans as "dendrites" and humans think of them as the input point for information coming in from another office. Once memos have been placed in Kevin's mailboxes, they will be processed by Kevin at his desk. Humans call this desk the "cell body" (also known as the "soma"). Remember, at the center of the desk (cell body) is Kevin (the nucleus). Kevin has an important job. At this point, his job is to make decisions about the memos that came from Stephanie. For example, Kevin must decide how significant and urgent the information contained in the memos is. As a side, Kevin also has additional responsibilities such as keeping his office functional, keeping it organized, etc. In human terms, a nucleus is responsible for keeping the neuron functional, is involved in synthesizing proteins, and helps to store genetic information of the neuron.

Now, let's say Kevin has processed Stephanie's memo and begins typing a response to it. If Kevin determines that this situation is *significant*, he will send his typed document to his office printer to be printed. This sending of the document to his printer is known to humans as an "action potential." It refers to an electrical signal that fires based on a combination of present ions (such as sodium and potassium). It is important to note that, for Kevin, he either clicks "print" on his computer screen or not. There is no in-between. He cannot "sort of" print a document. It is also important to note that once he sends a document to be printed, there is short period of time where he cannot print again (the refractory period). Once this refractory period is over, however, Kevin will be able to send another document to be printed. Now, Kevin's computer and his printer are connected by a wire. Humans term this wire an "axon." Axons carry electrical information to the next destination. In our example, the next destination is Kevin's printer. Many axons are insulated to speed up transmission. Since Kevin's document is very important, its speedy transmission is also quite important. Therefore, many of these wires are insulated and to humans, this insulation is known as the "myelin sheath." The fatty myelin sheath helps to speed up an electrical impulse. You will also notice in Figure 2.1 that there are gaps between the myelin sheath that humans term the "Nodes of Ranvier." These gaps help to boost the speed of transmission even more!

So, at this point, Kevin has taken Stephanie's memo, processed it, made some decisions, and determined it to be quite important. Important enough to create his own documented response to it and print it. At this point, he has clicked "print", and the document has been transmitted. Since this is such an important document, however, Kevin's office has several printers all set up to print at once. To humans, these printers are known as the "axon terminals." Now, his document has been printed numerous times, placed in special envelopes (known to humans as "vesicles") and are now considered memos (neurotransmitters) ready for distribution. Next, Kevin's memos need to go somewhere. As seen in Figure 2.3, the envelopes full of his memos are ready for distribution. In human lingo, neurotransmitters will be released from vesicles and will enter the "synapse" which is a gap or space between Kevin's office (pre-synaptic neuron) and Ron, his colleague's office (post-synaptic neuron). It is essentially the hallway between offices and now, Kevin is getting ready to send this information to Ron. Once the memos are ready to be taken out of the envelope, they will be placed in the correct mailboxes (post-synaptic receptors) for Ron. Ron will then process Kevin's memo and the information will continue down the line to other offices. Ron will also send any extra memos back to Kevin (this is known as "reuptake" to humans), and they can either be shredded or placed back into envelopes for future use and documentation.

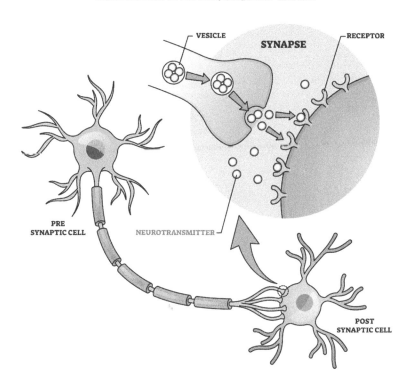

Figure 2.3 - Memo Distribution

Figure 2.4 is also a rough schematic of the offices in our example. Keep in mind that there is variability in the processes here at NeuroVoy. Not all offices look and function the same. For example, some offices have more mailboxes, and some have smaller desks. In human lingo, this could mean some neurons have more dendrites or have

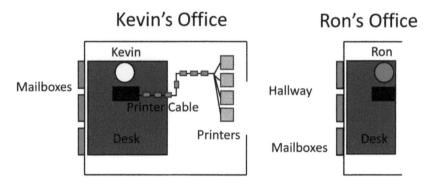

Figure 2.4 - Office Schematic

smaller somas. Another important point is that offices are broadly divided into two types. We have grey offices (grey matter) and white offices (white matter). The office designations make it more clear about some of the functions that their employees are responsible for. For example, employees that work in white offices have printer cables with insulation (myelinated axons), so they are more involved in communication. Employees that work in grey offices do not have this insulation and are more involved in information processing. However, the overall mission of our company is the same: to do the best we can for our clients. All day, every day. Below is a list that could be helpful in further understanding the two terminologies.

Employee = Cell nucleus

Office = Neuron

Memo = Neurotransmitter

Mailboxes = Dendrites

Desk with laptop = Cell body (Soma)

Printer cable = Axon

Printer cable insulation = Myelin sheath

Printer cable speed boosters = Nodes of Ranvier

Electrical signal from computer to printer after clicking "print" = Action potential

Period when "print" cannot be clicked. Signal cannot be sent during this time = Refractory period

Printers = Axon terminals

Security envelope = Vesicle

Hallway = Synapse

Neighboring mailbox = Post-synaptic receptor

Extra memos sent back to original office = Reuptake

Support Staff You May Encounter

We also provide a number of support staff for our primary employees so that they may be able to effectively function. To humans, our support staff are "cell nuclei" who also have workstations known as "glial cells." There is a roughly similar number of support staff workstations for each primary office. There are numerous support staff, however, broadly we will discuss four main support staff groups. These groups include Maintenance, Information Technology (IT), Security, and Fluid Management (FM).

The first support staff group we would like to introduce you to is Maintenance. To humans, these support staff (cell nuclei) are found in workstations known as "astrocytes." These staff are extremely important to the job being done here at NeuroVoy and their workstations are highly specialized. One of our top maintenance crew members, Charlie, and his workstation can be seen in Figure 2.5. Notice the star-like shape. This is typical of these workstations. There are a few things that we want to make you aware of in terms of what our maintenance crew are responsible for. Please keep in mind, this is not an exhaustive list and for more information on their full list of responsibilities, please contact us at DER or refer to your supervisor. First, our maintenance crew and their workstations are involved in helping to maintain and repair our offices, including the physical structures such as the walls and the lights. This is important as you could imagine problems with the office could prevent Kevin from communicating with other employees, such as Ron and his office. In human lingo, astrocytes are important in repair and maintaining the physical structure of and communication between neurons. Our maintenance crew and workstations are also involved in helping to maintain optimal working conditions for our offices. For example, for Kevin to be most productive, the

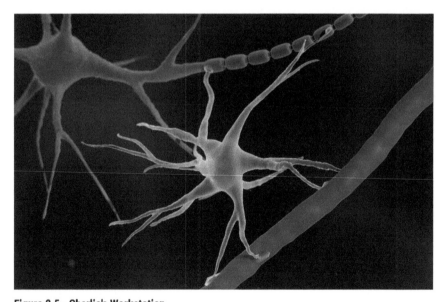

Figure 2.5 - Charlie's Workstation

conditions of his office need to be comfortable. In human lingo, astrocytes help to regulate pH and ion levels to keep a balance or "homeostasis" in the CNS. The maintenance crew and workstations are also involved in keeping the supply room stocked with snacks for our primary employees. We get it! Our employees get hungry, and we want to keep them happy so that they remain productive. In human lingo, astrocytes manage the availability of essential nutrients (such as glucose) for neurons to make sure the neuron, including the nucleus, has energy and can function properly.

Our next group of support staff we would like to make you aware of is Information Technology (IT). To humans, these support staff (cell nuclei) are found in specialized workstations called "oligodendrocytes" which can be seen in Figure 2.6. Our IT staff have several primary responsibilities but one of the most important includes the installation

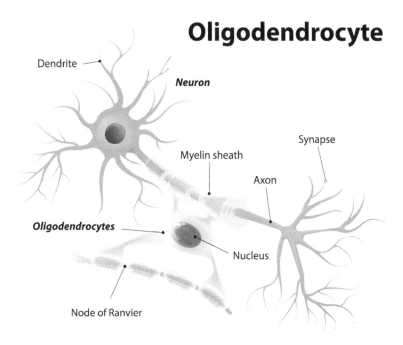

Figure 2.6 - An IT Workstation

and maintenance of the printer cable insulation. In human lingo, oligodendrocytes help to create and maintain the myelin sheath that surrounds an axon. As a side note, Periphalink Solutions also has

IT staff and workstations (to humans, they are known as "Schwann cells").

Our third group of support staff we would like to make you aware of is Security. To humans, these support staff (cell nuclei) are found in specialized workstations called "microglia" and can be seen in Figure 2.7. Our security team is essential to protecting our primary employees, support staff, and building. In human terminology, microglia are involved in the immune response in the central nervous system. Our security team quickly responds to potential threats, but they are also involved in some secondary responsibilities. For example, we do not want our employees and other staff to get hurt while they do their job, so, our security team is also tasked with clearing any debris or waste they might encounter while on patrol. Further, when they encounter

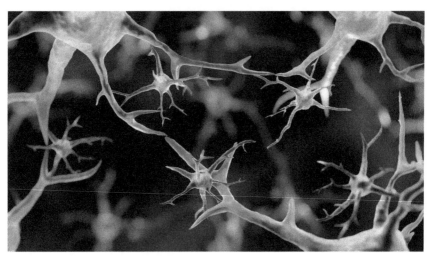

Figure 2.7 - Security Hard at Work

damage within the building, they will set out to repair and restore it. As you can see, they function as security, but you can almost think of them as "security plus."

Finally, we would like to introduce you to our Fluid Management (FM) staff. To humans, these staff (cell nuclei) are found in specialized workstations called "ependymal cells" and can be seen in Figure 2.8.

Figure 2.8 - Fluid Management

These support staff and their workstations are involved in creating the fluid within the building and then managing the flow. In human lingo, ependymal cells are involved in the creation of "cerebrospinal fluid" and then helping to circulate it. In the next section of the handbook, we will discuss the building fluid in more detail.

Your Office and Role

We have now introduced you to the main components and general processes of an office and additional support staff you may encounter. We would now like to turn your attention to your office and role, specifically, here at NeuroVoy Industries. As mentioned in the previous section of this handbook, you are joining an already established company through a process called neurogenesis. It is again important to note that our employees and offices are specially organized within our company by their abilities. Put another way, neurons are organized by the central nervous system based on characteristics such as anatomy and function.

So where will you be working?

You will be working within the Hippocampal Team of the Temporal Department in the Left Zone of our building. In human lingo, you will be located within the left hippocampus (part of the temporal lobe of the cerebral cortex). You are part of a very specialized subteam known as the Dentate Subteam (dentate gyrus). One of your primary roles on this team is to **help form and store memories for humans**. We will talk about teams, departments, committees, and more later on in this handbook. Due to location and function, your office (granule cell) is set up slightly different than some of the offices we have discussed thus far (like Kevin's). This set up is extremely important, however. For example, Dentate Subteam offices are smaller but placed very close together and have many mailboxes to allow for maximum communication and productivity. Put another way, the cell body of granule cells are smaller with many dendrites. You will be receiving more information about day-to-day responsibilities from your Team Leader, Jennifer.

| 3 |

Our Building

Now that you have been oriented to the general processes and offices within our company, we would like to shift your focus to the layout of our building. Although our headquarters (Figure 3.1) and navigating it may appear extremely complex to newcomers, the layout is actually beautifully and intelligently designed (in our opinion!). It is a place where our primary employees and support staff can work seamlessly and creatively together. It is designed to encourage teamwork while offering all of the necessary equipment and amenities to help promote productivity.

Again, at the core of our mission is committing to putting our clients first. They entrust us to get the job done and the functionality and design of our building allows us to do that! This portion of the handbook is set up to discuss our building in sections. It will probably be most useful to start from the "bottom-up" in terms of the layout. In the next section of the handbook, we will discuss different departments and teams in more detail and where they are housed. From there, we will discuss committees that span across different areas within our building, departments, and teams. This section, however, is a crucial step in the understanding of how these departments and committees function.

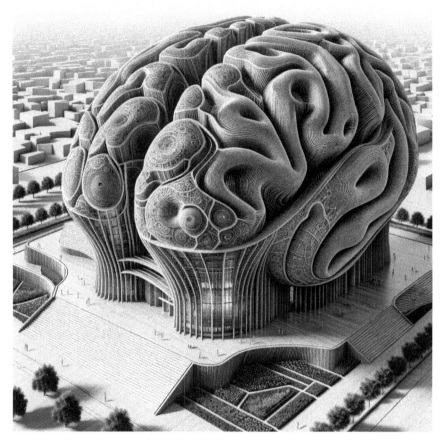

Figure 3.1 - NeuroVoy Industries Headquarters

It is important to note that our building is divided symmetrically into a Left Zone and Right Zone (left and right hemispheres of the brain). You can see this very clearly in the image above with the Left Zone (left hemisphere) being more prominently displayed. Understanding that there are two zones is important because some departments and teams are housed in both zones. In contrast, for some tasks and functions, one zone is considered to be more "dominant" for our clients and more likely to house the department or team that will engage in that task or function. This will be discussed further in the handbook. What you cannot see clearly in this image but that is equally important is our Central Tower and Elevator System. This is where we will start before moving our way up to the top levels.

Elevator System

We will start with the Elevator System to begin our journey through our building. To humans, this system is known as "spinal cord" which can be seen quite nicely in Figure 3.2. A unique feature of this system is simply how far underground it extends (into the human body). Another unique property of this part of our building is that it is home, exclusively, to the Spinal Cord Division of NeuroVoy Industries. The Elevator System is called this because massive amounts of information flow through this system via what are known as "tracts." There are numerous elevators that run through this system, and it is important to note that, depending on the function, information travels up (ascending tracts) or down (descending tracts) and can stop at

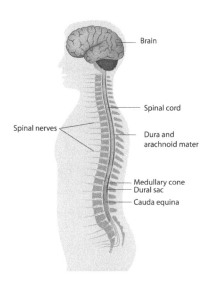

Figure 3.2 - The Elevator System Prominently Displayed

various levels. This can be seen in Figure 3.3, which is a cross section of our system. Our upward elevators take sensory information from the outside human world and pass this information along to the Brain

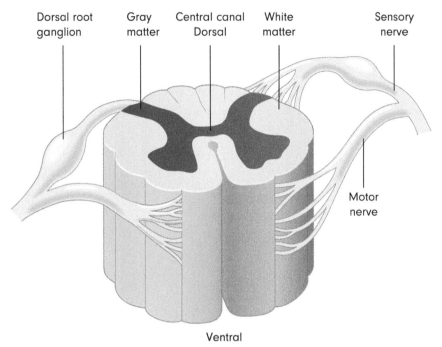

Figure 3.3 - Cross Section of Elevator System

Division of NeuroVoy Industries. There are various elevators including some that help humans process what is known as "proprioception." This refers to the perception and awareness of their body movement and position. Specific elevators involved include the "dorsal column-medial lemniscus pathway" and "spinocerebellar tract." Another elevator helps to process pain and temperature (spinothalamic tract). Our downward elevators take information from the Brain Division and send along to various parts of the human body. Some are involved with voluntary motor control (corticospinal tract), balance and posture (vestibulospinal tract), and the coordination of movement (rubrospinal tract). Though thoughtfully designed, we will later discuss potential problems that may occur at this level of our building. The Elevator System connects with the lowest level of the Central Tower, which is discussed in the next section of the handbook.

The Central Tower

Figure 3.4 displays a nice image of the Lower Level of the Central Tower. The Central Tower is broadly divided into the Lower Level and Upper Level. The Lower Level (known to humans as the "brainstem") of the Central Tower connects to the Elevator System and houses some very important departments including the Medulla Department, the Pons Department, and the Midbrain Department. Collectively, these departments are responsible for sending and receiving information to and from the Elevator System and to and from the Upper Level of the Central Tower. They are also involved in basic survival functions for our clients, such as regulating their blood pressure, respiration, and more. These departments will be discussed further in the next section of the handbook.

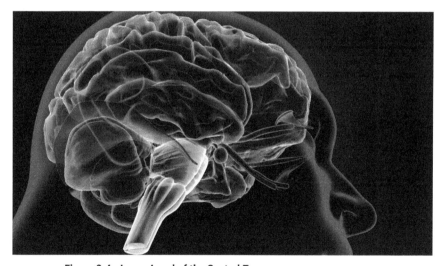

Figure 3.4 - Lower Level of the Central Tower

The Upper Level of the Central Tower, as seen in Figure 3.5, primarily includes the Thalamic Department, Hypothalamic Department, Cingulate Department, Pituitary Department, and the BG Department. These departments are responsible for a wide variety of tasks and functions that include involvement in the relaying of sensory information,

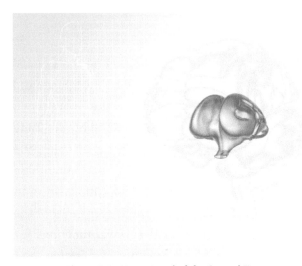

Figure 3.5 - Upper Level of the Central Tower

homeostasis, movement, stress response, emotional processing, and decision making. The Central Tower also has office space adjacent to the Lower Level known as the Central Tower Annex. This area is home exclusively to the Cerebellum Department which deals primarily with voluntary movement and balance. The location of the Central Tower

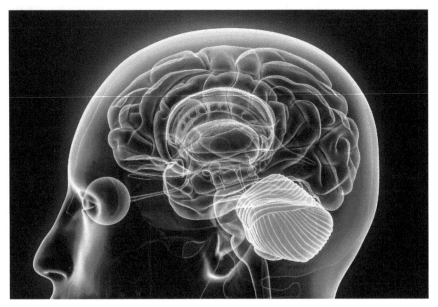

Figure 3.6 - The Central Tower Annex

Annex can be seen in Figure 3.6 and humans will sometimes call this area the "Little Brain" because of the similarity to the setup of the Outer Wings (discussed next).

The Outer Wings

Outside of the Central Tower, the other primary office spaces are located in the Outer Wings of the building. Humans refer to these wings of the building as "lobes" and they are collectively known as the "cerebral cortex." There are four Outer Wings in total: the Frontal, Temporal, Parietal, and Occipital Wings (lobes). Remember, our building is also divided, broadly, into two zones (left and right hemispheres). So, for example, we have a Left Frontal Wing (left frontal lobe) and Right Frontal Wing (right frontal lobe). This is prominently displayed in Figure 3.1 at the beginning of this section of the handbook. The Outer Wings are home to departments that deal with completing higher order tasks for our clients. For example, these departments are involved in memory, language, executive functioning, attention, visuospatial functioning, and more. In Figure 3.7, you can see where the

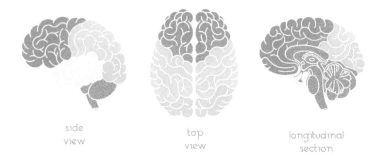

Figure 3.7 - Various Views of the Outer Wings

wings are designated. They are color coded with the Frontal Wing in orange, the Temporal Wing in yellow, the Occipital Wing in green, and the Parietal Wing in blue. Similar to the other areas within our building, we will explore their departments and teams in more detail in the next section of the handbook. In Figure 3.7, you can also see various angles or views of the Outer Wings. The leftmost image shows the "side" or "sagittal" view of the Outer Wings. This is the viewpoint that will be used to discuss most departments (but not all) within these areas later in the handbook. The middle image shows all wings minus the Temporal Wing from an aerial view. The rightmost image shows how the Outer Wings surround the Central Tower. On the underside of the Frontal Wings is where the Olfactory Teams work. This location is known as the Olfactory Expressway (the olfactory bulbs). You will learn about this team in the next section of the handbook.

Skyways and Bridges

Within our building, there are several notable connectors that help to bridge large areas together. Humans might think of this as akin to a

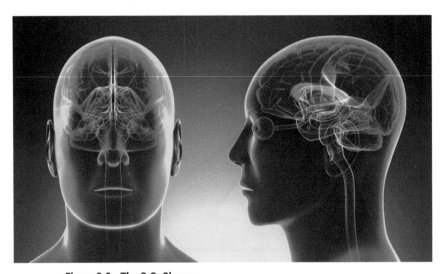

Figure 3.8 - The C.C. Skyway

skyway or bridge. One of the largest of these bridges is what we call the C.C. Skyway (the corpus callosum) which can be seen in Figure 3.8. This massive skyway connects the Left Zone (left hemisphere) and the Right Zone (right hemisphere) of the building. This allows for smooth transmission of information between departments and teams on each side of the building. The AF Bridge (the arcuate fasciculus) is another major connector within our building and helps to transmit information between some of our departments that are involved in speech and language. You will be oriented to these departments in the next

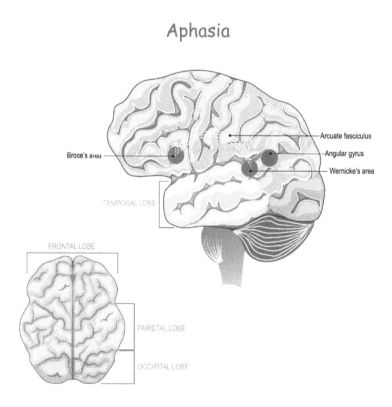

Figure 3.9 - Aphasia Poster and Location of the AF Bridge

section of the handbook, but to illustrate the location of the AF bridge, we have provided an aphasia poster in Figure 3.9. This was developed

28 - DEREK A. MCKAY

by our Language Committee for our clients. Aphasia will also be discussed later in the handbook. There are certainly other bridges within our building, but these are the two that we wanted to orient you to.

Additional Building Information

Beyond the Elevator System, Central Tower, and Outer Wings, there are some additional and important considerations to the structure and function of our building. Due to the nature of the work we do here at

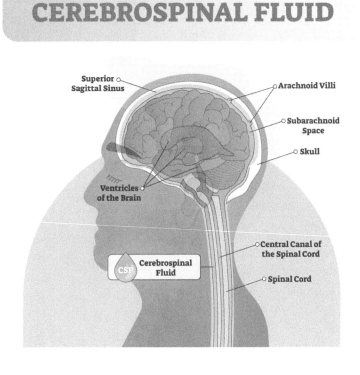

Figure 3.10 - Cerebrospinal Fluid System

NeuroVoy, our building must be protected. Therefore, there are some layers that surround our building that are not observable in Figure 3.1. First, our building has a series of waterways akin to "plumbing" in a human house. These waterways flow outside of our building and are filled with what is known as "cerebrospinal fluid (CSF)." CSF also flows through the building via what are known to humans as "ventricles." This can be seen in Figure 3.10. CSF provides several major benefits to our clients. First, our building is surrounded by it thereby providing some basic protection. Further, similar to plumbing, it works to remove waste from our building and to help create a stable environment. CSF is continually generated by specialized employees that are collectively known as the Choroid Plexus Department (the choroid plexus). Outside of the CSF, other primary layers that humans discuss at length are the "meninges" and the "skull." The skull is the large, global protector of our building whereas the meninges work in different layers to provide additional support. These are seen in Figure 3.11 which was a promotional poster we made for our client when he was in high school. It's an oldie but a goodie.

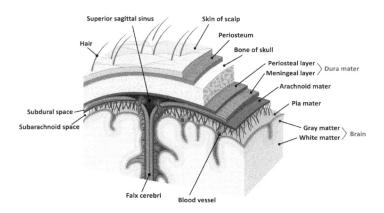

Figure 3.11 - Protective Layering Promotional Poster

Humans refer to the three layers of the meninges as the "dura mater, arachnoid mater, and pia mater." The dura mater is the outer layer and the toughest of the three. The arachnoid mater provides cushion, but is not as tough as the dura mater. Finally, the pia mater is thin and delicate, but covers our building completely and directly. NeuroVoy's protective layering is very important but just like anything else, there can be complications. For example, sometimes our clients experience what humans term "tumors." These abnormal growths can have significant effects on our building because we do not have the space to allow them to grow. This can put pressure on and destroy other parts of our building. We will discuss this in a later portion of the handbook. Another very important aspect of our building is what humans call the "cerebrovascular system." This system exists around and through our building and can be seen beautifully in Figure 3.12. To humans, this system includes arteries, veins, and capillaries that provide oxygen and nutrients to different areas of the brain and spinal cord. This is akin to the electrical system in a human house that allows lights, appliances, heating and cooling systems, and more to function properly through cables and wires. You can imagine if the electricity to the kitchen stopped working in a human house (and especially for an extended period of time), this could greatly impact the refrigerator and all of the food inside of it. This is similar to NeuroVoy. Imagine one of our departments, such as the Medulla Department, losing all ability to function properly. Unfortunately, this does happen for various reasons which we will discuss later in the handbook. Keep in mind that the size of the cables and wires in this system varies. For example, very thin wires exist in a human house such as those that allow humans to charge their cell phones. This would be akin to some of the smaller arteries in the brain. Bigger appliances, though, such as a refrigerator or washing machine have much bigger and thicker wires. This could be considered similar to the larger arteries in the brain. Larger arteries include the Anterior Cerebral Arteries (ACAs), Middle Cerebral Arteries (MCAs), and Posterior Cerebral Arteries (PCAs). Further, going back to Figure

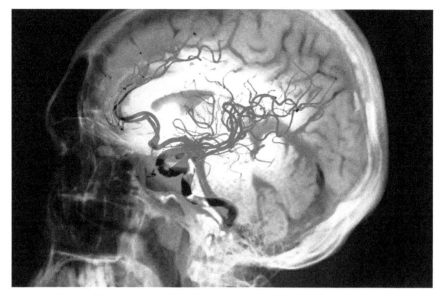

Figure 3.12 - The Cerebrovascular System

3.11, issues can arise that cause blood to be released between the different protective layers (known to humans as a "hemorrhage" such as a "subarachnoid hemorrhage") which, again, is significant as there is really no place for the blood to go.

Lastly, we would like to direct your attention to another specialized system that originates at various levels of the Central Tower (the nuclei originate here) but work largely within the confines of Periphalink Solutions (peripheral nervous system). We call this system the "Automation System" whereas humans call this system the "cranial nerves." In total, there are twelve pairs of cranial nerves, and they assist our clients in numerous automatic sensory and motor functions. For example, some cranial nerves help move our clients' eyes, manage their balance, or turn their necks. Therefore, we often liken this to the automation systems in a human house. For example, this is akin to advanced technology and sensors that control different functions such as with lights, heating, entertainment, security devices, and more. Figure 3.13 on the following page displays the Automation System in more detail. The twelve "individual systems" with their corresponding numbers are as follow:

Olfactory (1), Optic (2), Oculomotor (3), Trochlear (4), Trigeminal (5), Abducens (6), Facial (7), Vestibulocochlear (8), Glossopharyngeal (9), Vagus (10), Spinal Accessory (11), and Hypoglossal (12).

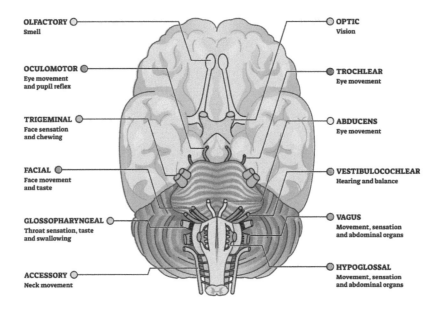

Figure 3.13 - Automation System

| 4 |

Major Departments and Teams

Now that you have a basic understanding of our building, we would like to introduce you to our major departments and, in some cases, specific teams within a department. For the purposes of this handbook, we will be outlining the major departments. Please keep in mind that this is not an exhaustive list of them all. If you do need more information about specific departments or teams that are not listed, we recommended discussing further with your supervisor or us here at DER.

Medulla Department

The Medulla Department (the medulla) is housed in the Lower Level of the Central Tower. As you may recall from the previous section of the handbook, this set of offices connects directly with the Elevator System. This is seen nicely in Figure 4.1, and it is important to note, information traveling up and down from the Elevator System also travels through the Medulla Department. In this way, the Medulla Department acts as a conduit for sending information between the Elevator System and other departments, such as those in the Upper Level of the Central Tower and the Outer Wings. This department also houses portions of the Automation System (cranial nerves 9 through 12) and is involved in basic life functions and regulating involuntary functions for our clients. This includes helping to regulate their heartbeat, breathing, blood pressure, and some reflexes such as swallowing and vomiting. It is also a member of the Reticular Formation Committee which will be discussed in the next section of the handbook.

Pons Department

The Pons Department (the pons) is also extremely crucial to the basic survival of our clients. As a reminder, the Pons Department is also located in the Lower Level of the Central Tower and also serves as a

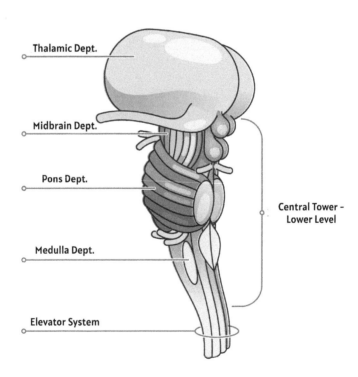

Figure 4.1 - Elevator System and Various Central Tower Departments

conduit of information between the Medulla Department and the Midbrain Department. Similar to their neighbors, the Pons Department also houses portions of the Automation System (cranial nerves 5 through 8) and is involved in life support functions for our clients. For example, the Pons Department is involved in respiration and helps to keep our clients alert and aroused (not sexually, per se) as part of the Reticular Formation Committee. The Pons Department can also be seen in Figure 4.1.

Midbrain Department

As you know, the Midbrain Department (the midbrain) is also located in the Lower Level of the Central Tower. Broadly, the Midbrain Department is involved in movement, reward, pain interpretation, and more. This department is also part of the Reticular Formation Committee, and we would like to orient you to some specific teams within this department including:

- The Roof Team (humans call this team the "tectum") is housed at the "top" of the Midbrain Department. It is home to even more specialized subteams including the Colliculi Subteam (humans refer to this as the combination of the inferior colliculus and superior colliculus which are involved in automatic auditory and visual attention, respectively). For example, if our client is at home and hears fireworks in the neighborhood or is at a baseball game and a foul ball is hit in their direction, the Colliculi Subteam helps to *automatically* orient them to those stimuli.

- The Tegmen-Team (humans refer to this team as the "tegmentum") houses portions of the Automation System (cranial nerves 3 and 4). It is also home to the VTA Subteam (the ventral tegmental area) which works quite closely with dopamine and is part of the Reward Committee (discussed in the next section of the handbook).

- The SN Team (humans refer to this team as the "substantia nigra") is involved in helping our clients move and is also involved in reward. This team works substantially with dopamine and is a member of the Reward Committee. The Midbrain Department also shares the SN Team with the BG Department.

Cerebellar Department

The Cerebellar Department (the cerebellum), located in the Central

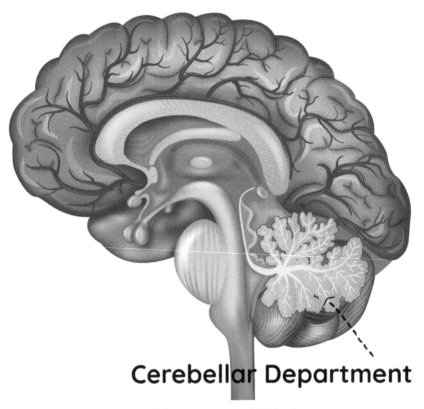

Figure 4.2 - The Cerebellum Department Highlighted

Tower Annex, is mostly known for their work with helping our clients to move. However, this department is also responsible for some other very important things. Figure 4.2 is a great image of the inner workings of the department and as you remember, the Central Tower Annex is called the "Little Brain" by humans. You can see why quite nicely in this image. Despite being most known for their involvement in the movement of our clients, they are also involved in learning and memory, especially related to what humans call "nondeclarative" or "implicit memory." An example of this would be the procedural memory of how to ride a bicycle or to drive a car. It would be quite difficult for our clients to learn how to drive **every** time they got in their vehicle and so this department helps our client to remember the procedure of driving. So, as you can see, this department's responsibilities are quite expansive and their office space within the overall context of our building is quite large!

Thalamic Department

The Thalamic Department (the thalamus) can be seen in Figure 4.1 and is located in the Upper Level of the Central Tower. The Thalamic Department functions as our central mailroom (or relay station) and is primarily responsible for taking information in and sending it out to where it needs to go. Humans sometimes refer to the thalamus as an airport. For example, airplanes may be flying into Cleveland from Miami. That plane will then either return back to Miami or fly off to another destination, such as Las Vegas or Los Angeles. This is essentially what this department does, however, instead of airplanes, they process primarily sensory and motor information.

Hypothalamic Department

The Hypothalamic Department (the hypothalamus) is a small, but mighty department located in the Upper Level of the Central Tower which is nicely observed in Figure 4.3. It is involved in directing many of our clients' "maintenance" activities and attempting to keep their body at homeostasis (or balanced). This includes, but is not limited to regulating hunger, thirst, sexual arousal, temperature, and fear. The Hypothalamic Department is part of the Limbic and Neuroendocrine Committees.

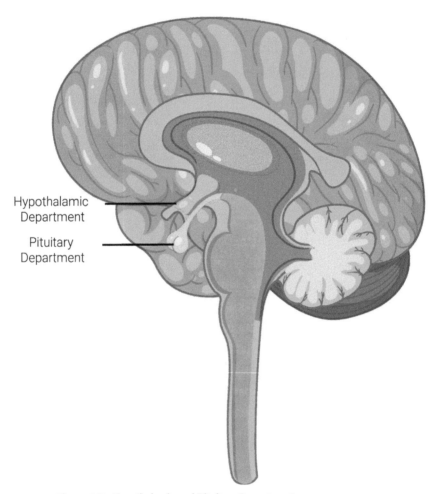

Figure 4.3 - Hypothalamic and Pituitary Departments

Pituitary Department

The Pituitary Department (the pituitary gland) is also located in the Upper Level of the Central Tower. This department serves as the

President of all Glandular & Organ Departments (directs the activity other glands and organs throughout the human body) and is a member of the Neuroendocrine Committee, similar to the Hypothalamic Department. In fact, the Hypothalamic and Pituitary Departments work quite closely with each other. Our clients' bodies and homeostasis are highly regulated by the work of the Pituitary Department. This department can also be seen in Figure 4.3.

Cingulate Department

The Cingulate Department (the cingulate cortex) is located in our two zones within the uppermost portion of the Upper Level of the Central Tower and can be seen in Figure 4.4. This department is involved in various cognitive functions of our clients such as decision-making. It

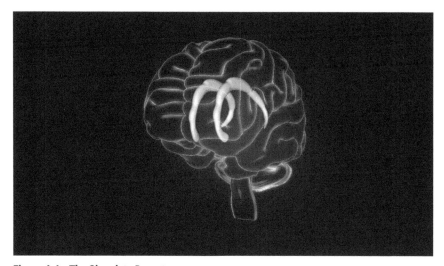

Figure 4.4 - The Cingulate Department

is also involved in emotion formation and processing. Interestingly, not only does this department receives incoming information about our clients' physical pain, but it also processes emotional pain such as that from being bullied or ostracized. The Cingulate Department is a member of the Limbic Committee and is implicated in many mental health diagnoses.

BG Department

The BG Department (the basal ganglia) is also located in both the Left and Right Zones (although rather closely together) and includes the SN team described prior. The BG Department also includes the CN Team (the caudate nucleus), GP Team (the globus pallidus), the NA Team (the nucleus accumbens), the Putamen Team (the putamen), and the Subthalamic Team (the subthalamic nucleus). This department, broadly, is involved in helping in the motor control of our clients and

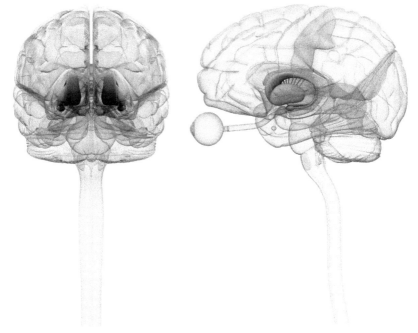

Figure 4.5 - The BG Department

working as an intermediary between other departments such as the Frontal Lobe Department, the Parietal Lobe Department, the Thalamic Department, and the Cerebellar Department. Many of the teams within the BG Departments act as an input point or an output point for motor information. This department is also implicated in procedural learning and memory, eye movements, emotion, and cognition and is part of the Limbic Committee and Reward Committees. Figure 4.5 indicates the location of the BG Department.

Frontal Lobe Department

The Frontal Lobe Department (the frontal lobes) is located in the "front" of the Outer Wings of our building in the Frontal Wings. To humans, this lobe of the cortex is found around both eyes, forehead, and the temples on both sides of the head. This department is responsible for quite a few things, and we have found that it is helpful to orient you to their major teams including:

- The PFC Team (the prefrontal cortex), located in both zones of our building, which provides our clients with numerous abilities. For example, this team helps our clients to not only plan out behaviors but also to follow through with completing them. This team is also heavily involved, broadly, in conducting higher order cognitive abilities such as memory and executive functioning. The latter includes things such as helping our clients to solve problems, reason through and make decisions, make sound judgements, organize things, and so on. This team is also involved in our clients' personality and mood and has three primary subteams. These subteams include the Orbitofrontal Subteam (the orbitofrontal prefrontal cortex), Dorsolateral Subteam (the dorsolateral prefrontal cortex), and Medial Prefrontal Subteam (the medial prefrontal cortex). The Medial Prefrontal Subteam also works closely with the Cingulate Department (specifically, the anterior portion of the cingulate cortex). We will discuss complications to each subteam in a later portion of the handbook. You can see the location of the PFC Team in Figure 4.6.

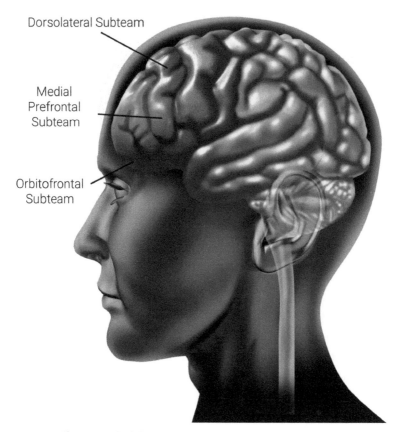

Figure 4.6 - PFC Team

- The Broca Team (Broca's area) is primarily involved in coordinating the production of speech and language for our clients and can be seen in Figure 4.7. This team is a major contributor to the Language Committee, discussed more in the next section of the handbook. It is primarily found in one zone of our building.

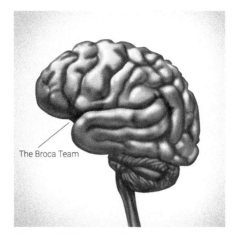

Figure 4.7 - The Broca Team

- The FEF team (the frontal eye fields) assists with our clients' voluntary eye movements and helps to create gaze. They can be located in Figure 4.8. and are found in both zones of our building.

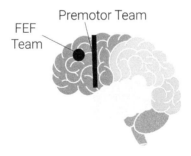

Figure 4.8 - FEF and Premotor Teams

- The Premotor Team (the premotor cortex) works to help our clients **sequence** movement and to make new sequences more automatic. Their approximate location can also be seen in Figure 4.8. For example, when our clients first learn to drive a car, the sequence of movements such as bucking the seatbelt, turning on the engine, checking their mirrors, putting the car in drive (or reverse), pressing the gas pedal, pressing the brake pedal, moving the steering wheel, using the turn signals, etc. is not very familiar. Over time, however, the sequencing of these steps becomes more automatic, and the Premotor Team helps with this. This team works closely with the Motor Strip Team (seen in Figure 4.9) described next. The Premotor Team is housed just in front of the Motor Strip Team in both zones of our building.

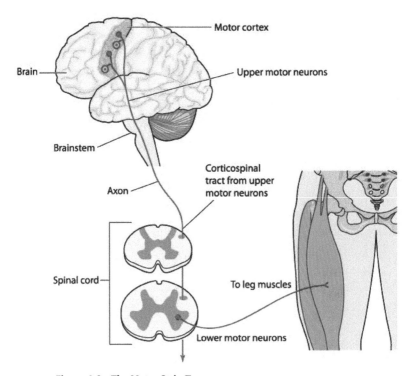

Figure 4.9 - The Motor Strip Team

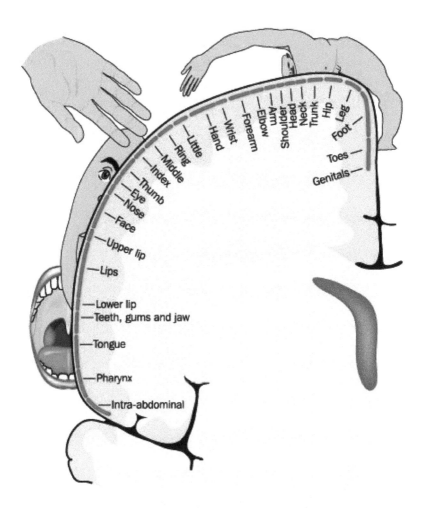

Figure 4.10 - The Homunculus

- The Motor Strip Team (the primary motor cortex) is found in both zones and is quite involved in the strength and coordination of movement for our clients. Members of the Motor Strip Team are responsible for different parts of our clients' bodies such that our client's arm is represented by specific Motor Strip Team members, for example. Humans call this representation the "homunculus." Body areas that require more strength and coordination receive a larger portion of the homunculus. A version of the Somatosensory Team's homunculus and can be seen in Figure 4.10. Lastly, you will notice in Figure 4.9 that the Motor Strip Team of the Left Zone will eventually help to coordinate movement and strength in the right leg. This is a concept known as "contralaterality."

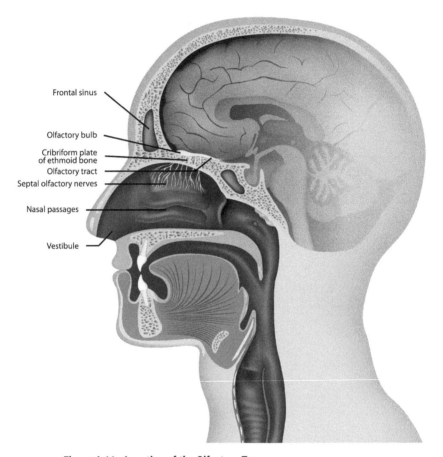

Figure 4.11 - Location of the Olfactory Teams

- The Olfactory Teams are actually located within the Olfactory Expressways (the olfactory bulbs) and helps to transmit olfactory (smell) information to be processed by other departments and teams. You can see the location of the Olfactory Teams in Figure 4.11.

The Temporal Lobe Department

The Temporal Lobe Department (the temporal lobes) is located in the "sides" of the Outer Wings of our building in the Temporal Wings. To humans, this lobe of the cortex is found around their temples and ears on both sides of their head. This department is also responsible for many things, and we would like to introduce you to the major teams found within this department including:

- The Hippocampal Team (the hippocampus), which is involved in helping our clients convert short-term memories into long-term memories and is a member of the Limbic and Reward Committees. This team is also involved in spatial navigation which, as an example, refers to our clients' ability to go from their house to the local park without use of a GPS or their cell phone and relying solely on their memory. Please note that the Hippocampal Team exists in the two zones within our building and can be seen in Figure 4.12. If you recall, this is the team that you will be working on and more specifically, you will be working on the specialized Dentate Subteam.

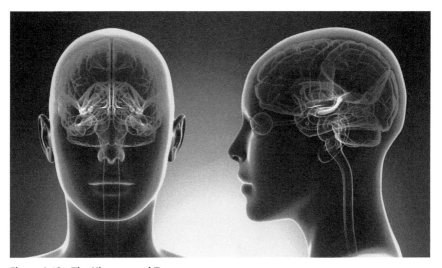

Figure 4.12 - The Hippocampal Team

- The Amygdalar Team (the amygdala) works closely with the Hippocampal Team to process emotion such as fear, anxiety, and aggression and to help form memories for our clients. This team is also involved in interpreting social cues. Similar to and working next door to the Hippocampal Team, the Amygdalar Team exists in both zones and can be seen in Figure 4.13. This team is also a member of the Limbic and Reward Committees.

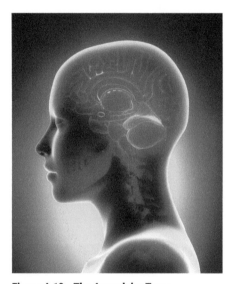

Figure 4.13 - The Amygdalar Team

- The Auditory Team (the auditory cortex) is found in both zones of our building and close to our clients' ears. In fact, the information that is obtained by our clients' ears is transmitted to the Auditory Team, which is the final stop in the processing of this information, as seen in Figure 4.14. This team not only processes sound but also pitch and the direction that the sound is coming from.

WELCOME TO NEUROVOY INDUSTRIES - 49

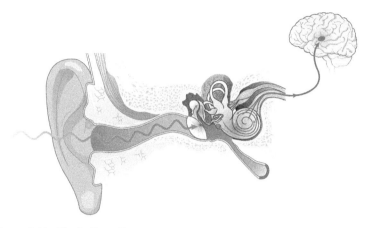

Figure 4.14 - The Auditory Team

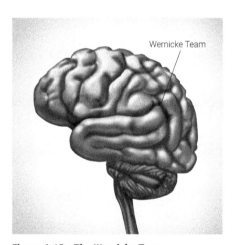

Figure 4.15 - The Wernicke Team

- The Wernicke Team (Wernicke's area) is involved in language comprehension for our clients and is a major contributor to the Language Committee. This team works quite closely with the Broca Team of the Frontal Lobe Department. The Wernicke Team can be seen in Figure 4.15 and is typically only found in one zone of our building.

The Occipital Lobe Department

The Occipital Lobe Department (the occipital lobes) is found in the "rear" of our building in the Occipital Wings and is primarily known for its role in visual processing. This department has many contributors (V1 through V6), but the V1 Team (the primary visual cortex) serves

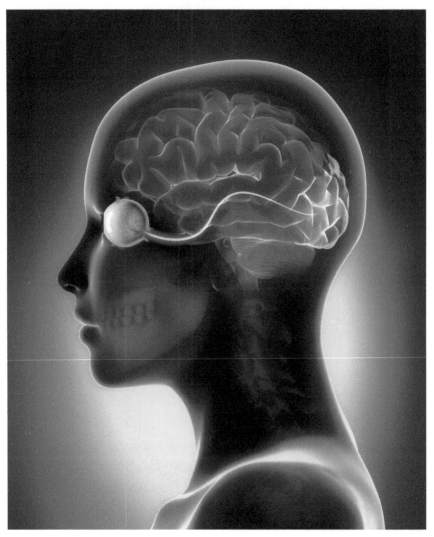

Figure 4.16 - The V1 Team

as a primary team within this department and can be seen in Figure 4.16. This team helps our clients to process visual information, such as shapes, colors, motion, depth perception, etc. This team helps to send information to important output points via two pathways. These pathways are known to humans as the "ventral" and "dorsal streams." The ventral stream (also known as the "what pathway") ends in the Temporal Lobe Department and helps our clients understand "what" something is. The dorsal stream, on the other hand, is known as the "where pathway" and helps our clients understand where something is located in space and if it is moving, for example. Dorsal stream information ends up in the Parietal Lobe Department. So, imagine our client has just seen a cat. The ventral stream will help the client understand that what he just saw was, in fact, a cat. The dorsal stream will indicate to the client that the cat is moving (perhaps across the windowsill).

The Parietal Lobe Department

The Parietal Lobe Department (the parietal lobes) is located in the "upper portion and to the rear" of the Outer Wings of our building in the Parietal Wings. To humans, this lobe of the cortex is found around the top of the head extending towards the back of the head. This department is involved primarily with sensory integration, and we would like to introduce you to some of their teams including:

- The Somatosensory Team (the somatosensory cortex), which is quite involved in the processing of sensory information for our clients. This includes the processing of touch, pain, temperature, etc. Similar to the Motor Strip Team in the Frontal Lobe Department, members of the Somatosensory Team are responsible for the sensory processing of different parts of our clients' bodies. Meaning, for example, touch for our client's fingers is represented by specific team members. Similarly, our client's lips are represented by other team members. Body areas that require more tactile (touch) information get a larger portion of the homunculus such as in Figure 4.10. You can see the location of one half (in the Left Zone) of the Somatosensory Team in Figure 4.17.

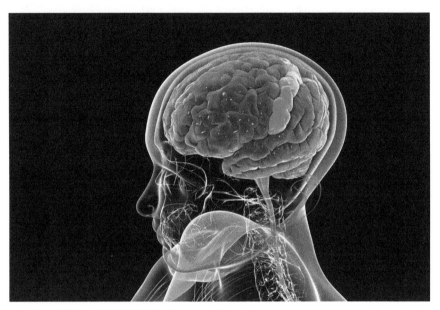

Figure 4.17 - The Somatosensory Team

- The Angular Team (the angular gyrus), seen in figure 4.18, is part of the Language Committee and is involved in reading and processing of written information for our clients. They are also involved in processing mathematical operations, copying/drawing, and more. This team is located in both the Right Zone and Left Zone.

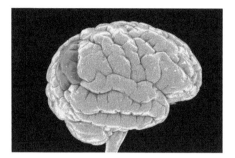

Figure 4.18 - The Angular Gyrus Team

- The Supra Team (the supramarginal gyrus), seen in figure 4.19, is part of the Language Committee and, similarly to the Angular Team, is also involved in the processing of written information such as assigning meaning to words. This team is also located in both the Right Zone and Left Zone. In terms of location, these two teams are very close to each other. However, they can be differentiated where the Angular Team is near the junction of the Parietal, Occipital, and Temporal Departments and the Supra Team is situated just in front of the Angular Team.

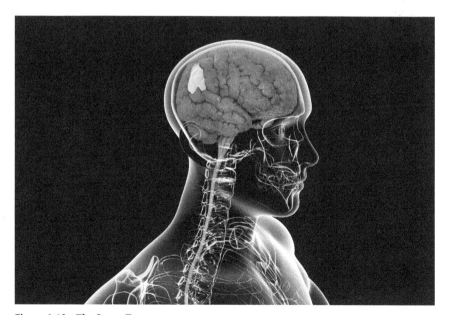

Figure 4.19 - The Supra Team

Insular Department

The Insular Department (the insular cortex) is found in an interesting location on both sides of our building. It is a bit "hidden" and found at the juncture of the Frontal, Parietal, and Temporal Wings. This department is involved in our clients' self-awareness, social-emotional processing, homeostasis, motor control, and visceral reactions to stimuli such as the feeling of "disgust." The Insular Department can be seen in Figure 4.20.

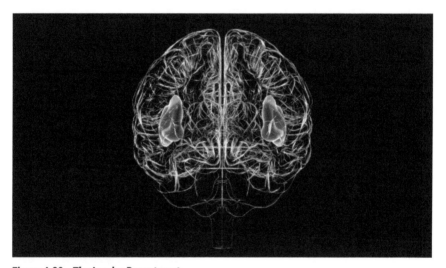

Figure 4.20 - The Insular Department

As a reminder, these are some of the primary departments and teams, but certainly not all of them. Should you have more in-depth questions about other teams, for example, please reach out to your supervisor or DER.

| 5 |

Important Committees

You have now been introduced to some of the major departments and teams within NeuroVoy Industries. From here, we would like to orient you to some important committees that exist within our company. Many of these committees have already been referenced throughout the handbook, but this is our chance to expand on them. Specifically, in this portion of the handbook, we will discuss five major committees including the:

- Reticular Formation Committee
- Limbic Committee
- Neuroendocrine Committee
- Reward Committee
- Language Committee

Again, to make it clear, there are other committees that are formed within the company, but these are the ones that we feel most necessary to describe during your onboarding process. Committees can be broadly thought of as a collection of departments and teams that, in themselves, do not have the necessary capability to complete certain functions. As a group, however, they do. There can be, of course, complications with our committees due to various factors that will be discussed in the next section of the handbook. However, let's begin with our first committee.

Reticular Formation Committee

The Reticular Formation Committee (the reticular formation) includes contributions from a number of departments including the Medulla Department, the Pons Department, and the Midbrain Department. Here is the official company policy on the purpose of this committee:

> "The charge of the Reticular Formation Committee is to oversee the vital functions of our clients related to arousal, attention, and sleep-wake cycles."

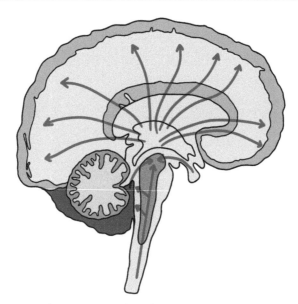

Figure 5.1 - Information Sent Through and by the Reticular Formation Committee

The Reticular Formation Committee is truly an integral network within our company. You can see in Figure 5.1 how information is first sent through the aforementioned departments and then dispersed throughout our company, for example, through the Thalamic Department and to the Outer Wings. A client scenario that the Reticular Formation Committee could be implicated in would be a college student waking up and going to class. This committee aids in controlling the student's sleep-wake cycle so that they wake up in time to get ready for class. This committee ensures that the student maintains arousal throughout the day and pays attention in class. Of course, there are things that can interfere with this committee's ability to do their job, some of which are described in the next section of the handbook.

Limbic Committee

The Limbic Committee (the limbic system) includes contributions from a massive network within our company including the Cingulate Department, the Hypothalamic Department, the Amygdalar and Hippocampal Teams of the Temporal Lobe Department, and more specified contributions from the BG Department and the PFC Team of the Frontal Lobe Department. The official company policy on the purpose of the Limbic Committee is as follows:

> "The charge of the Limbic Committee is to oversee the vital functions of our clients related to emotion, behavior, and memory."

Figure 5.2 shows the collection of members of the Limbic Committee. A client scenario that the Limbic Committee could be implicated in would be a child walking to school on the first day of classes. On the way, they pass by a house that has a large, barking dog in the yard. The dog cannot get to the child, but follows the child along the chain-link fence and barks loudly as the child walks on the sidewalk. The Limbic Committee may work together to identify emotions such as fear and anxiety related to this scenario. Let's say the child takes the same route

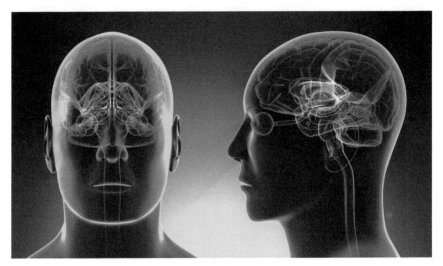

Figure 5.2 - Limbic Committee Members

tomorrow and the same dog continues to follow the child and bark. Again, this is processed by our committee as fear and anxiety. Over time, this committee will send this information to long-term memory storage so that the child realizes that that the dog will most likely be at that house along the route. It may also aid the child in choosing a new route that does not involve the dog, thus reducing fear and anxiety. Given this information has been sent to long-term storage, the child recalls not to take that specific route to school every day for the entire school year. Again, this is just one possible scenario that the Limbic Committee is involved in. However, you should be able to recognize the clear integration of the departments and teams listed. Similar to the Reticular Formation Committee, there are things that can interfere with the production and ability of the Limbic Committee that we will discuss in the next section of the handbook.

Neuroendocrine Committee

Before we discuss the Neuroendocrine Committee, it is important to note that this committee actually works with a company outside of NeuroVoy Industries. This company is known as Endocrine

Enterprises (the endocrine system). Endocrine Enterprises consists of numerous factories (glands) that produce and release chemical packages (hormones) for our clients (throughout the body). Endocrine Enterprises is crucial in regulating many processes for our client. However, NeuroVoy Industries is equally crucial. The Hypothalamic Department and the Pituitary Department are the two primary, involved committee members from our company as they help to determine what the Endocrine Enterprises factories are producing, how much they are producing, when they are releasing the chemical packages, etc. In certain functions, Periphalink Solutions offers support through their Autonomic Division (especially through their Sympathetic Department). Here is the official company policy on the purpose of this committee:

> "The charge of the Neuroendocrine Committee is to oversee the vital functions of our clients related to homeostasis, metabolism, stress response, and reproduction."

Before we talk about the major NeuroVoy Departments involved in this committee, we would also like to discuss what humans term the "SAM Axis" which stands for Sympatho-Adreno-Medullary Axis. The SAM Axis is activated during an initial threat or response to stress. Think of this as the rapid response to stress. Remember in the first section of the handbook when we discussed those two hungry lions waiting for our client? Surely, in this scenario, the SAM Axis would take the lead. This would include contributions from the Sympathetic Department of Periphalink Solutions and chemical packages from the Medulla division of the Adrenal Factory (a portion of the Adrenal Gland which sits atop the kidneys and not to be confused with the Medulla of the brain). But, in this scenario, let's assume that the incident with the lions is not a quick one. Meaning, maybe our client lives in a remote area where help cannot get to them quickly. So, instead of the lions being captured and returned to the zoo quickly, maybe our client is now trapped in their house for a week while the lions claw on the windows and try to break down the door. In this scenario, a different collection of committee members may eventually take over in what humans term the "HPA Axis" and includes the Hypothalamic and Pituitary Departments as noted above but also the specific factories such as the Adrenal Factories. These interactions occur as a delayed response to stress (in contrast to the SAM Axis) and ultimately results in the release of a chemical package known as "cortisol" (sometimes referred to as the "stress hormone" by humans). Cortisol can be quite useful to our clients in regulating stress response, blood pressure, inflammation

STRESS RESPONSE

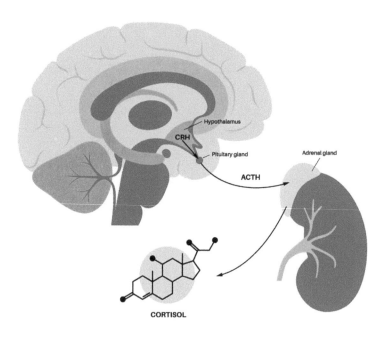

Figure 5.3 - The HPA Axis within the Neuroendocrine Committee

and so on. Too much cortisol, though, can be damaging to our company and to the client as a whole. For example, too much cortisol has been linked to reduced immune functioning and susceptibility to illness, cardiovascular disease, changes in mood, and more. The HPA Axis can be seen in Figure 5.3 which is a promotional poster for our clients. In it, you can see the hypothalamus, pituitary gland, adrenal gland, and cortisol represented. You can also see CRH which stands for "corticotropin-releasing hormone" and ACTH which stands for "adrenocorticotropic hormone." CRH is produced by the Hypothalamic Department which stimulates the production of ACTH by the Pituitary Department. ACTH will eventually make its way to the Adrenal Factory which signals the factory to begin producing cortisol.

Reward Committee

The next committee we would like to direct your attention to is known as the Reward Committee (mesolimbic and mesocortical systems). Here is the official company policy on the purpose of this committee:

> "The charge of the Reward Committee is to oversee the vital functions of our clients related to reward, motivation, and learning."

This committee is led primarily by contributions from the VTA Subteam of the Midbrain Department, the NA Team of the BG Department, the PFC Team of the Frontal Lobe Department and the Cerebellar, Hippocampal, and Amygdalar Departments. Figure 5.4 displays, for example, the transmission of dopamine from the VTA Subteam to other areas such as the NA Team and PFC Team. This committee helps our clients to want to repeat things that help them to live (and to make it easier to do those things again) such as drinking water. So, for example, if our client is very thirsty and is offered a glass of water, specific members of the committee determine that, if their client drinks the water, they will no longer be thirsty. Once they drink some water,

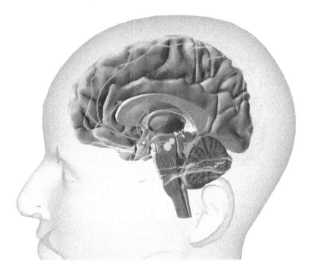

Figure 5.4 - Reward Committee Processes

these committee members determine that this satiating and dopamine is released. This information is sent to other members of the committee that help to form memories that this water was delicious and refreshing (and perhaps who gave the client the water). In turn, it makes it more likely that they drink water when they are thirsty and potentially seek that water out from a specific person or place (e.g., maybe the client stops at a convenience store to get water every morning before going to the gym). Again, this committee is a powerhouse, but occasionally can be interfered with by other forces such as substance use by the client. We will discuss this more in the next section of the handbook.

Language Committee

The Language Committee includes contributions from a number of teams and departments including the Broca Team of the Frontal Lobe Department, the Wernicke Team of the Temporal Lobe Department, and the Angular and Supra Teams of the Parietal Department. The AF Bridge is also heavily utilized by this committee to transmit information. For a refresher on the location of much of the Language

Committee, please refer back to Figure 3.9. Here is the official company policy on the purpose of this Committee:

> "The charge of the Language Committee is to oversee the vital functions of our clients related to language production and reception."

Please keep in mind that this committee, depending on the client, is involved in any of the following: spoken language, written language, or signed language. A scenario in which this committee would be highly involved in for our client would be ordering a meal at a local fast-food restaurant. Once our client enters the restaurant and assuming visual ability is intact, the Angular and Supra Teams process written information for what is on the menu (e.g., the #1 meal is a hamburger, french fries, and a drink for $12.99). Our client then begins to interact with the restaurant employee to place their order. Our client perceives the speech produced by the employee ("May I take your order?") which is then interpreted by the Wernicke Team as meaningful information. The Broca Team then allows our client to produce the speech needed to respond to the employee ("I would like the #1, please."). Again, the AF Bridge acts as a major pathway between the Broca and Wernicke Teams and would theoretically allow our client to repeat what the employee said, if needed. As stated before, there are certainly additional committees that function within our company and have significant roles, however, we have limited the discussion to these ones in the handbook. In the next section, we will discuss potential complications that can happen to our building, our departments and teams, our committees and of course, to our clients.

| 6 |

Potential Complications

Our company strives to do the best we can for our clients. Occasionally, however, errors and complications do occur often by no fault of our own. There are many factors that can be related to these complications and the purpose of this portion of the handbook is to discuss them in further detail as we want you to be aware and prepared. To begin, we must remind you that our company functions as an ecosystem with employees, support staff, and our building working together. One of our mottos is "Keep Alive and Help Thrive" meaning, our overall goal is to not only keep our clients alive and functioning properly, but to also help them thrive in their lives, relationships, jobs, etc. There are a number of reasons, however, why our company may not be able to do this effectively (or, at all, in some cases) such as damage to our building or disruption to departments, teams, and committees. This handbook will certainly not cover all of the possible reasons (etiologies) for why there may be complications, but we will focus on a few broad possibilities.

First, as we discussed in section three of the handbook, our building has a series of waterways akin to "plumbing" in a human house. CSF helps to create a stable environment and remove waste from our building. A major problem, however, is that this system can become "plugged up" so to speak. Similar to pipes in a human house, there may be a clog in the Cerebrospinal Fluid System that causes imbalance in the system and difficulty removing waste. This can result in a condition known as "hydrocephalus" in which the ventricles in the brain become enlarged and can be seen in Figure 6.1. Again, think of water backing up in the pipes of a house. For our clients, this may mean changes in their thinking, walking, and mood. Hydrocephalus can be congenital (from birth) or acquired later on in life. The primary treatment for hydrocephalus includes surgery to place a shunt which will help to drain the CSF properly.

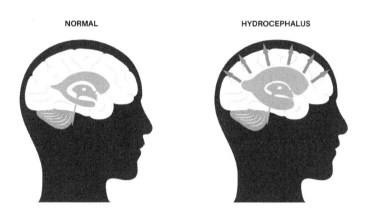

Figure 6.1 - Hydrocephalus

Remember earlier discussions about what humans call "the cerebrovascular system" which includes arteries, veins, and capillaries that provide oxygen and nutrients to different areas of the brain and spinal cord? As you know, this system is similar to the electrical system in a human house that allows lights, appliances, heating and cooling systems, and more to function properly. Well, those things in the human house may continue to work (but improperly) or stop working at all. In our world, we may have entire departments or teams that may no

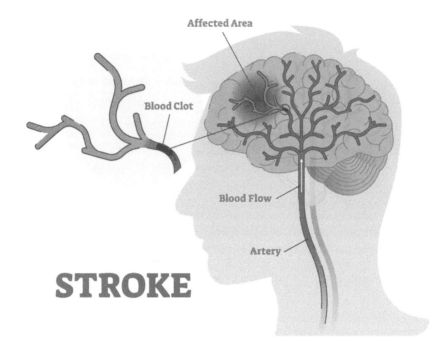

Figure 6.2 - Example of Ischemic Stroke

longer have the ability to work properly or again, may function in part, but not incompletely. One of the most common reasons for this is what humans term a "stroke." Strokes can happen for all sorts of reasons but are usually characterized by a lack of blood supply (ischemic stroke) or bleeding into the brain tissue (hemorrhagic stroke). Depending on the location of a stroke, the departments, teams, and committees in those areas may be impacted. An example of an ischemic stroke can be seen in Figure 6.2. The extent to which they are impacted depends on a number of things, including the location and severity of the stroke, potential recovery post-stroke, etc. So, to reiterate, in our human house analogy, a stroke could exist in the kitchen of a house which could effectively kill electricity to all areas of the kitchen or impact only a few appliances. In this case, it is possible that other areas in the house are working just fine. Again, this is very dependent on the location and significance of the stroke. For example, we will discuss this later, but a stroke to the Lower Level of the Central Tower could effectively cause the entire house to shut down. Stroke also puts our clients at a higher risk for developing conditions later on, such as a vascular dementia.

Some of the other etiologies that we will reference *could be* an abnormal growth on or within our building (tumor; Figure 6.3), damage from the elements or some external force (trauma; Figure 6.4), our building eroding and losing structural integrity (degenerative conditions such as Alzheimer's disease; Figure 6.5), our equipment not working properly such as damage to communication between our computers and printers (multiple sclerosis; Figure 6.6), or disruptions in or improper balance of memos that impact the function of certain departments and teams (many neurological and mental health conditions).

Figure 6.3 shows an image from a bottom-up view of an MRI (known as an axial view). You'll notice a "white" mass towards the

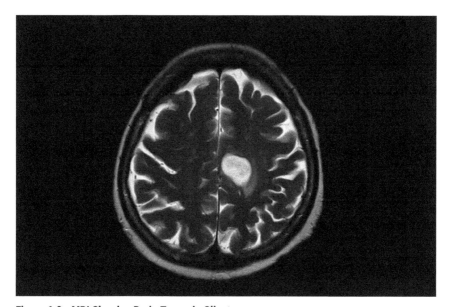

Figure 6.3 - MRI Showing Brain Tumor in Client

center of the image. This is an abnormal growth on our building which can unfortunately damage it which impacts the departments and teams in that area. In human lingo, this is a brain tumor that is likely impacting the client's cognition and possibly motor/sensory function. Please keep in mind that the specific issues will be related to where the abnormal growth is and, in turn, what departments and teams are implicated. Prognosis varies and in some cases, NeuroVoy Industries will no longer be able to function due to the passing away of our client. Treatment may include surgical removal of the growth, radiation, chemotherapy, or combination therapy (although these treatments can have unwanted effects, too). Mental health treatment may also be indicated depending on changes in mood, increase in mental health symptomology, adjustment to diagnosis, etc. Figure 6.4 shows another axial view of a client's brain. As you can see, the Protective Layering has been damaged causing damage to Frontal and Temporal Wings of the Left Zone (on neuroimaging, left equals right and vice versa). In human lingo, this client has sustained a traumatic brain injury (TBI)

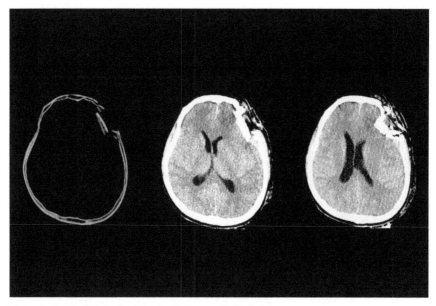

Figure 6.4 - CT of Brain for Client with Traumatic Brain Injury

that has caused a skull fracture and swelling to the left frontotemporal region. The departments and teams housed in this area may experience a temporary disruption in productivity and, if significant enough, prolonged disruptions. Treatment may involve surgery, mental health treatment, and different therapies such as speech therapy, physical therapy, occupational therapy, and/or vestibular therapy depending on the severity and extent of the damage. Figure 6.5 shows an image of the deterioration or eroding of our building. Think of this as a destruction of the integrity of our building. In human lingo, this is a coronal view (as if you are looking at the client's face) of an MRI of a brain for a client with presumed Alzheimer's disease. Alzheimer's disease is a form of dementia characterized by memory impairment and often impacts the Hippocampal Team. Widespread decline in cognition will generally occur to the point where the person is no longer able to take care of themselves on their own. One of the hallmark signs of an Alzheimer's disease brain is the presence of what are called "amyloid plaques" and "neurofibrillary tangles." Basically, these plaques and tangles are abnormal clumps of different kinds of proteins that eventually lead to loss of neurons (building blocks of the brain) and atrophy (shrinking of the brain and associated enlarged ventricles) which can be seen in Figure 6.5. Low levels of the memo (neurotransmitter) acetylcholine are often

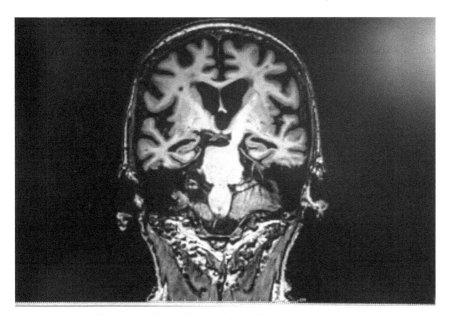

Figure 6.5 - MRI of Brain for Client with Alzheimer's Disease

also implicated. We are still unsure (fully) as to why some people develop this disease. At current moment, however, there is no cure. Some clients may benefit from medications that work to increase the level of acetylcholine, however, those benefits are often short-lived. Also keep in mind that there are more conditions that can lead to the eroding of our building (e.g., frontotemporal dementia). If you recall from section one of the handbook, we discussed our employee Kevin and his computer. His computer is connected to his printer by a cable (axon) which carries electrical information. Also recall that this cable is insulated (myelinated) to speed transmission of information.

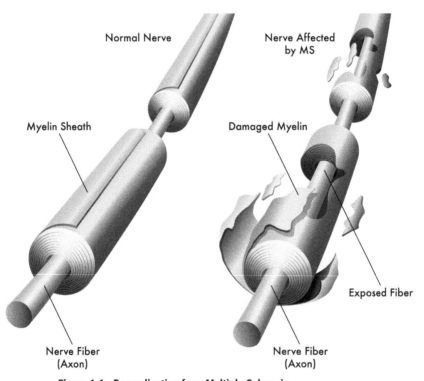

Figure 6.6 - **Demyelination from Multiple Sclerosis**

Figure 6.6 is an illustration of what can happen to our printer cables and insulation when damaged by certain conditions. This damage can cause the communication between our computers and printers to become slowed or to not happen at all. In this instance, the specific condition observed in Figure 6.6 is an autoimmune disorder known to humans as "Multiple Sclerosis" or "MS." MS, like a stroke, can impact numerous departments and teams (sometimes simultaneously). Therefore, for one client, symptoms may include cognitive decline, motor impairment, sensory impairment, and more. For a different client with MS, only motor impairment may be impacted, for example. Unfortunately, there is no cure for MS but, humans can use what are known as "disease modifying treatments (DMTs)" to help slow the progression. In our world, that would mean the destruction of our printer cables and insulation happen at a less rapid rate. Keep in mind, however, once they are destroyed, they are destroyed.

Elevator System

The first specific area that we would like to discuss as a potential complication is damage to our Elevator System (the spinal cord) and the Spinal Cord Division of NeuroVoy Industries. If you recall from earlier in the handbook, information travels via numerous tracts within this system. Therefore, damage to these tracts can result in a spinal cord injury, which, depending on the tract could impact:

- Voluntary movement
- Coordination of movement
- Balance and posture
- Understanding of body position and movement
- Pain and temperature processing

Obviously, depending on the tract or tracts that are damaged, our clients may have notable disruption in those areas, which can ultimately impact their day-to-day functioning. Damage to our elevator system can also impact information being sent to other areas of our building and ultimately to departments and teams within the company. There are numerous reasons why this system may be damaged including trauma (such as if our client gets into a car accident), cancer, infections, and more. The extent to which this system may be repaired depends on the extent of the damage and the location. Sometimes our clients engage in rehabilitation or may need surgery to repair this system, and this may help some clients. For other clients, however, the damage to this system may be permanent. Figure 6.7 displays some of the possible outcomes from a spinal cord injury.

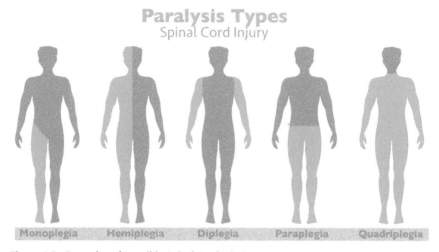

Figure 6.7 - Examples of Possible Spinal Cord Injuries

Lower Level of the Central Tower

The Lower Level of the Central Tower includes the Medulla Department, the Pons Department, and the Midbrain Department and are involved in basic survival functions for our clients, such as regulating their blood pressure, heart rate, respiration, and more. Therefore, damage to these areas can be especially life-threatening. Their involvement in the Reticular Formation Committee could also mean disruptions in attention/arousal (such as being in a coma). Some humans note, "the real estate is prime" in this location for these very reasons. For example, imagine a stroke occurring in this area and impacting one or more of these departments. If we go back to section four of the handbook, we can surmise what *might* happen if these departments or teams are not functioning properly. Keep in mind, these assumptions are just that. They are assumptions. Meaning, our client could potentially suffer damage to the Lower Level of the Central Tower, but may not display a symptom that we would expect clinically. Let's walk through some of the implications of a department or team that may not be functioning properly.

First, since the Medulla Department helps regulate our clients' heartbeat, breathing, blood pressure, swallowing, and vomiting, complicating factors, such as a stroke, could cause difficulties and problems with any one of these symptoms and ultimately lead to the shutdown of our entire company (death). Sensory and motor information travel to and from the Elevator System could also be disrupted, as well, as arousal, attention, and sleep-wake cycles due to this department's role in the Reticular Formation Committee. Similar complications could also occur with issues to the Pons Department with additional problems such as with respiration and facial/eye movements through the Automation System. In rare cases, our clients could experience what is known as "Locked-In Syndrome" due to complications with the Pons Department. In this situation, our clients are fully paralyzed, but are still conscious, retain their cognitive abilities, and may only be able to communicate through eye movements. Complications to the Midbrain Department could also potentially result in difficulties with reward, orientation to stimuli, movement issues, transmitting pain information, etc. The Automation System is also highly vulnerable with complications to this area of our building.

Central Tower Annex

Complications to the Central Tower Annex and the Cerebellar Department may impact our clients' ability to coordinate movement and balance. Ataxia (lack of voluntary movement) and/or difficulties with implicit memory may be present. There are numerous etiologies for problems with the Cerebellar Department including strokes, tumors, and multiple sclerosis. Temporary issues with this department are sometimes seen when our client ingests too much alcohol. It appears that alcohol significantly impacts our employees in the Cerebellar Department. In turn, our client may have difficulty walking (e.g., stumbling, lack of coordination, etc.). For example, a client driving erratically might get pulled over by the police and be asked to perform a sobriety test which involves their ability to walk heel-to-toe along a straight line. Given the importance of the cerebellum for coordinating these precise walking movements while maintaining balance, the police are essentially testing our Cerebellar Department employees' levels of activity which can be significantly compromised by the presence of alcohol. In other cases, damage to this department may be of a traumatic or biomedical nature (e.g., disease; stroke) and doctors may ask our clients to perform the heel-to-toe walking test, or maybe rapidly turn their hands over and back (i.e., palms facing down then turning palms up, then repeating) in a neurological work-up. Again, this is essentially testing our department's ability.

Upper Level of the Central Tower

As noted in previous sections of the handbook, the Upper Level of the Central Tower includes the Thalamic Department, Hypothalamic Department, Cingulate Department, Pituitary Department, and the BG Department. Complications to the Thalamic Department can be particularly devastating. Remember that one of the primary responsibilities of this department is to serve as a relay station for incoming and outgoing sensory/motor information. The result from damage to this area of our building and subsequently to our Thalamic Department is usually profound neurological impairment. For example, a thalamic stroke may result in what is known as "hemi-neglect." This would mean that our client may have a deficit in attention/awareness on the side opposite of the stroke, sensory changes, motor weakness, speech/language difficulties, and higher order cognitive impairment. Again, think of this department as an airport. If you shut down "Terminal B" of a large, international airport, no planes (information) can fly in and out (be transmitted). You could imagine that this would disrupt passengers, employees, pilots, and the flight industry as a whole.

Complications to the Hypothalamic Department can result in imbalances in many of our clients' maintenance activities and homeostasis. Again, this would include things like the regulation of hunger, thirst, temperature, fear, and sexual arousal. To illustrate, we would like to discuss a genetic condition that humans have identified known as "Prader-Willi syndrome." In this condition, in which symptoms are seen in early childhood, genetic abnormalities on chromosome 15 interfere with the normal functioning of the Hypothalamic Department, causing a client to have a constant, insatiable appetite. As a result, clients diagnosed with Prader-Willi syndrome are constantly hungry, eat excessively, and are at risk for developing numerous cardiovascular conditions such as diabetes and hypertension. As of now, there is no cure. Again, other complications can arise to this department, such as from a stroke, which puts our clients' homeostasis at risk. Complications to this department can also negatively impact the functioning and duties of broader committees, such as the Limbic and Neuroendocrine Committees.

If you recall, the Pituitary Department serves as the head of or President of all other Glandular and Organ Departments. Complications with this department can result in disruption or the abnormal functioning of glands and organs throughout our clients' bodies. As you also may recall, the Pituitary Department works closely with the Hypothalamic Department and so homeostasis may also be significantly impacted due to complications with this department. One example of a potential complication may result in what humans term "Cushing's syndrome." Symptoms of Cushing's syndrome can be seen in Figure 6.8. This condition can occur due to a pituitary adenoma (a benign tumor) that can cause excessive secretion of adrenocorticotropic hormone

(ACTH) and cortisol. Symptoms can include high blood pressure, fatigue, weight gain, changes in mood, a rounder face, and irregular menstruation. Again, this is just one example of what can happen when there are complications to this department. Other examples can include what humans call "hypothyroidism" in which the thyroid gland does not produce enough thyroid hormone or "acromegaly" in which certain parts of the human body grow excessively.

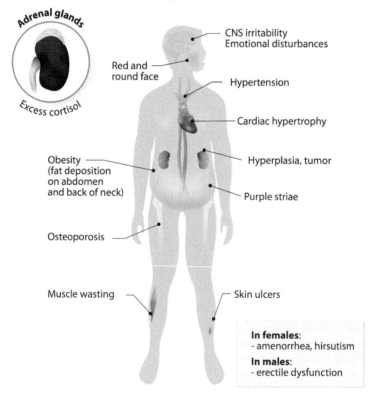

Figure 6.8 - Symptoms of Cushing's Syndrome

Complications with the Cingulate Department can have significant impacts on the Limbic Committee. For example, in our clients, this may include emotional and mood changes, changes in social behavior, or difficulty regulating emotion. Damage could also include alterations in pain perception or cognitive deficits, such as with decision-making. Potential etiologies for disruptions to this department can include stroke, trauma, tumor, infection, degenerative diseases, and mental health conditions. Lastly, complications to the BG Department may result in movement disorders such as Parkinson's disease or Huntington's disease given this department's role in working as an input/output point for motor information. Parkinson's disease is considered a hypokinetic disorder as it results in "less" movement such as rigidity, gait issues, tremor, etc. In this condition, our employees may not be able to utilize, transmit, or even produce certain memos (neurotransmitters) such as dopamine. That is, loss of dopamine is often related to the development of Parkinson's disease symptomology. Huntington's disease, rather, is considered a hyperkinetic disorder as it results in "more" movement including what is known as chorea. Chorea is

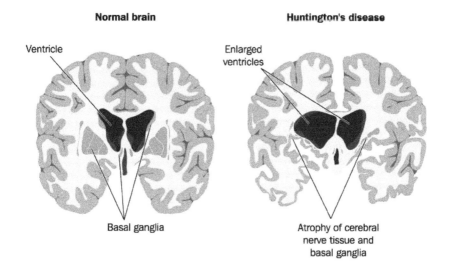

Figure 6.9 - Brain Changes in Huntington's Disease

characterized by involuntary, irregular movements that look "dance-like." Huntington's disease is often genetically inherited, and disruptions occur primarily to certain teams of the BG Department (Figure

6.9), such as the CN Team (caudate nucleus). In addition to these disorders, complications to the BG Department can also result in deficits in procedural learning and memory, eye movements, emotion, and cognition. Additional etiologies include stroke, for example. There is no cure for the aforementioned movement disorders.

Outer Wings

The Outer Wings are implicated in much of our clients' higher order thinking and cognition. Complications to these areas generally result in cognitive deficits such as with memory. However, complications can also exist with motor and sensory ability/interpretation, problem-solving, decision making, personality, mental health, language, processing speed, visuospatial functioning, and more depending on what departments and teams are affected. Let's start with the front of the Outer Wings which houses the Frontal Lobe Department and several teams.

A common etiology for disruption and issues to the PFC Team include traumatic brain injury. For example, from a fall, motor vehicle accident, or assault. The PFC team is in quite a vulnerable position within our building. Other possibilities include stroke, a degenerative condition such as frontotemporal dementia or Alzheimer's disease, a tumor, lesions from multiple sclerosis, and so on. Complications to this team may impact and alter our clients' memory, executive functioning, personality, mood, and more. Broad (diffuse) damage to this area of our building and to this team may result in problems with self-awareness. In extreme cases, our client may experience what is known as a "paramnesia." In these cases, disruptions in our processes here at NeuroVoy may result in the client being unable to distinguish between reality and fantasy. One such example is known as "reduplicative paramnesia" which refers to a delusion in which our client believes that a person or place has been duplicated and exists in two different places simultaneously. In a variant of this condition, "Capgrass syndrome" refers to our client believing that a person they know has been duplicated but that this person is an imposter. For example, our client may believe that their mother has been duplicated and this new duplication is an imposter mother. This is also observed in some mental health conditions such as schizophrenia. In other situations, different subteams of the PFC Team can be impacted. For example, complications to the Orbitofrontal Subteam can result in our clients appearing disorganized. They may have difficulty inhibiting themselves appropriately and regulating emotion. Complications to the Dorsolateral Subteam may result in our clients having difficulty with higher order cognition such as problem solving, judgment, decision-making, and memory. Complications to the Medial Prefrontal Subteam may impact our clients' motivation, attention, emotion, judgement, and even motor function to the lower extremities. Stroke, trauma, or neurodegenerative changes are common, but not exclusive, etiologies.

As mentioned before, the Broca Team is involved in the production and expression of language. The Broca Team is typically found in the "Main Zone" of our building. Humans refer to this as the "dominant hemisphere." All that really means is that the Broca Team is typically housed in one side of our building. For most people, this is the Left Zone, but not always. Complications to this team can result in what humans call a "Broca's aphasia." An aphasia refers to a disorder of language and in this particular example, our client would be able to understand language (assuming the Wernicke Team is functioning appropriately) but would not be able to produce it fluently. Sometimes this is called an "expressive aphasia." For example, a client with Broca's aphasia may **want to** say something like: "I need to go to the grocery store today to buy milk, eggs, sugar, and butter." However, when the speech is actually produced, it may sound very slow, halting, and nonfluent such as: "Need...store...err...buy...milk...err...egg...sugar...uh....butter." This same pattern is also observed in those who use sign language. One of the most common etiologies for Broca's aphasia is a stroke. Next, recall that the FEF Team helps our clients with voluntary eye movements and to create gaze. For example, this team helps the eyes move together, in the same direction. Complications can arise, however and produce difficulty shifting gaze, following moving objects, and coordinating eye movements. Trauma, infection, stroke, and tumor are potential etiologies for disruptions to the FEF team. The Premotor Team helps our clients to sequence their movements. Unfortunately, complications with this team may result in what is known as "apraxia." This occurs when a client is unable to perform a task or movement despite understanding what is being asked and having the physical ability to perform the task or movement. Typically, this also includes movements that our client previously knew. Complications with this team can also result in difficulties completing more complex motor sequences. Stroke, tumor, and neurodegenerative conditions are possible etiologies for complications to this team. The Motor Strip Team is often susceptible to stroke, as well. If you recall, members of this team are responsible sending motor signals through the spinal cord to be executed by the opposite side of the body and this is represented by the "Homunculus." For example, imagine there is a complication to our team members that work in the Right Zone and in the area of the Homunculus responsible for our client's arm. In this case, our client's left arm may be weak and they may have poor coordination. If a stroke is large enough, however, additional parts of the body may be impacted as well. Lastly, the Olfactory Team helps to transmit smell information to be processed by other departments and teams via the Olfactory Expressways. Sometimes, the Expressways can be damaged by trauma or neurodegenerative conditions which may impact the Olfactory Team's ability to transmit smell information properly. Our client may report not smelling things properly or weakly sensing the smell of something.

Next, let us move to the "sides" of the Outer Wings which houses the Temporal Lobe Department and several teams. We start first with the Hippocampal Team which, as you recall helps to consolidate short-term memories into long-term memories for our clients. Unfortunately, this team can be significantly impacted by neurodegenerative conditions, such as Alzheimer's disease or frontotemporal dementia. For example, if our client develops Alzheimer's disease, this team will likely have trouble functioning due to the erosion of this part of our building and subsequent shutdown of some or all of this department. Our client may be impacted by having memory difficulties such as remembering names, dates, appointments, to take medications, and more. Eventually, it will likely progress to the point of our client not being able to remember who people are and to lose the ability to take care of themselves independently. When the Amygdalar Team is negatively impacted, our client may have difficulty with emotional processing and memory formation, especially related to emotional memories. Recall that this team helps to process negative emotion (e.g., fear and anxiety) but they also help to processes positive emotion, like happiness, for learning and memory processes by the Hippocampal Team. In this sense, the Amygdalar Team helps to tell other teams that "this is important, we need to keep it for later." One issue, however, is that this team tends to want to keep both negative and positive emotionally-laden material. That can include things that are traumatic in nature. This team's work is also the basis of some conditioning processes and is highly involved in Limbic Committee processes as discussed in the last section of the handbook. Disruptions to this team can result in difficulty with emotional processing, social cues, and formation of memories especially related to emotional material. This team is implicated in numerous mental health etiologies as well such as diagnoses related to anxiety and mood. This could include major depressive disorder, generalized anxiety disorder, posttraumatic stress disorder, bipolar disorder, and more. In extreme cases, in which the teams in both zones of our building (e.g., both amygdalae) are damaged, we may see something that humans term "Kluver-Bucy syndrome." This is characterized by diminished fear response, low aggression, hyperorality (wanting to put things in your mouth), and hypersexuality. Possible etiologies include stroke, neurodegenerative conditions, or infection. It should be noted that the mental health diagnoses above are not solely due to disruptions to the Amygdalar Department and there are many other possibilities for the presence of these diagnoses.

Our Auditory Team is responsible for processing auditory information such as pitch and direction of the where the sound is coming from.

Disruptions to this team may result in the reduced ability or inability to process and be aware of sound (even if other areas such as the ears are working properly). Stroke, tumor, trauma, or neurodegenerative conditions are examples of possible etiologies can impact this team. As a side, recall the Colliculi Subteam? This subteam is involved in auditory and visual attention and the automatic orientation to stimuli. Imagine a client whose Auditory Team was unable to perform their duties, but their Colliculi Subteam was. So again, think about a client who is in a situation where maybe there are fireworks being set off in the neighborhood. If the Auditory Team is unable to function properly, but the Colliculi Subteam is, the client will be oriented to the sounds of the fireworks. However, they will not be able to processes the sound and therefore will be unaware of what the sound actually is.

Similar to the Broca Team, the Wernicke Team is found in the Main Zone and is involved in language. In this case, the reception of language. Complications to this team (usually due to stroke), can result in what humans call a "Wernicke's aphasia." In this example, our client would likely be able to produce language fluently, however, the actual language produced would not make sense. This is because the client would not be able to understand what language is actually being produced. Similarly, if asked questions, the client would respond with language that has no meaning, as they are unable to understand what is being asked. Using a similar example from before, a client with Wernicke's aphasia may be asked "Can you please go to the grocery store today and buy milk, eggs, sugar, and butter?" Since the client cannot fully understand what is being asked, they may produce a response such as, "Yes we are playing in the pool over there. We had a great time and I can't wait to drink more water in the sun." Again, this speech is quite fluent but has minimal-to-no meaning.

Next, we move to the major team within the Occipital Lobe Department in the "back" of the Outer Wings. The V1 Team is highly involved in the processing of visual information and, as you may recall from previous sections of the handbook, send information via two primary pathways (the ventral and dorsal streams). Also recall that there are other contributors to the visual process. We are simply focusing on the V1 Team. Complications to this team can result in a variety of difficulties for our client and stroke tend to be a major etiology for these difficulties. However, trauma, degenerative issues, and tumor are also possible etiologies. When there are complications to this team, our client may experience what is known as an "agnosia." An agnosia refers to a loss of ability to process sensory information and can be found in many forms (for example, a visual agnosia which is the loss of ability to recognize objects). It is important to note that the input point (such as the eyes) is not damaged in this scenario, rather, it is due to the impact on this particular team. In this type of agnosia, our client may be able to perceive a cat, but be unable to tell you that it is a cat. Depending on the severity and type of complication to this team, our client may have visual field defects (such as loss of visual field of one eye) or cortical blindness (intact, working eyes but the loss of interpretation for all visual stimuli).

In the "upper portion and to the rear" of the Outer Wings lies the Parietal Lobe Department. Complications to this department largely result in sensory interpretation and integration issues. For example, the Somatosensory Team helps to processes sensory information for our clients. When complications arise, our client may have difficulty processing and experiencing touch, pain, temperature, and more. Stroke is a common etiology that can results in disruptions to this team and, similar to the Motor Strip Team, Somatosensory Team members are responsibly for interpreting sensory signals from the opposite side of the body in the fashion of the "homunculus." For example, imagine there is a complication to our team members that work in the Left Zone and in the area of the homunculus responsible for our client's feet. In this case, sensory information to our client's right foot may not be interpreted correctly and our client may not report any feeling in their foot. If a stroke is large enough, however, additional parts of the body could be impacted as well, such as the leg. Similarly, strokes are primary culprits for impacting the Angular Team. If this team is impacted, our client may have difficulty with reading and processing written information. In some cases with the Left Zone Angular Team, a condition known as "Gerstmann syndrome" may develop. In this condition, our clients typically display a collection of four primary symptoms: inability to distinguish between left and right (left-right disorientation), inability to distinguish the fingers on a hand (finger agnosia), difficulty with or inability to write (dysgraphia/agraphia), and difficulty with or inability to learn/comprehend mathematics (dyscalculia/acalculia). With complications to the Supra Team, we may also see our clients struggle with processing written information and assigning meaning to words. As you can see, complications to this department can heavily impact the Language Committee.

Lastly, in a previous section of the handbook we mentioned an additional department known as the Insular Department that is found near the juncture of the Frontal, Parietal, and Temporal Wings and is often considered "hidden." Complications to this area can lead to increased apathy, difficulty with the perception of taste, altered social behavior and emotional processing, and more. Various etiologies exist for damage to this department, including stroke, trauma, infection, substance use, mental health diagnoses, and neurodegenerative disorders. We would also like to make a comment on some of the skyways and bridges referenced earlier. Damage to the C.C. Skyway can cause difficulty with our two zones communicating with each other. For example, teams in the Temporal Lobe Department of the Right Zone may not be able to communicate with teams in the Temporal Lobe Department of the Left Zone. Stroke is a common etiology. Surgical resection may also be a possible etiology. Sometimes this is done to prevent the spread of what humans call "seizures" from one side of our building to the other. By shutting down the C.C. Skyway, seizures are less able to spread. However, this may also result in what is called a "split-brain" in which our departments and teams in each zone may start to work independently of each other and not communicate. Put in human terms, this may cause the two hemispheres of the brain, and by extension both sides of the body to work independently of each other. Shutting down the C.C. Skyway is often a last resort. Additionally, the AF Bridge may become damaged (often from stroke) which disconnects Broca's team and Wernicke's team from communicating and working together. In human terms, this can cause what is known as a "conduction aphasia." This type of aphasia is characterized by fluent speech and intact comprehension but difficulty or the inability to repeat information. This aphasia is relatively rare, however.

| 7 |

Good Luck!

We hope that this handbook has provided you with an informational overview about the processes here at NeuroVoy Industries, our parent company, and associated companies. Throughout this handbook, you were exposed to offices, roles, and functions of the primary employees within our company, as well as, support staff that you may encounter. The layout of our building, major departments, teams, and major committees were also discussed. Lastly, even though we do our best and try to avoid them, potential complications that you, our company, our clients, and our building may encounter were mentioned. We must reiterate that this was not meant to be an exhaustive list of every single thing within our company, but, this should be a nice start for you as you begin your journey. Again, we are extremely excited for your arrival and are happy to have you as part of our organization. We here at DER are ready to assist with any concerns that you may have. Best of luck to you moving forward!

HUMAN GLOSSARY AND INDEX

The following glossary of terms is meant for humans, but again, it can sometimes be useful for our employees to understand their lingo. You will also see an accompanying page number where the term is either first discussed or heavily discussed in the handbook.

Acalculia (84) – The loss of or difficulty with ability to perform simple math computations usually due to an acquired injury or neurological condition.

Acetylcholine (8) – A neurotransmitter involved in many functions including muscle contraction, cognition, and sleep.

Acromegaly (76) – A condition in which the pituitary gland produces excess growth hormone which can cause enlargement of body systems.

Action potential (10) – An electrical impulse that travels along the axon of a neuron.

Adrenocorticotropic hormone (61) – A hormone that is produced by the pituitary gland that plays a role in the stress response by stimulating the adrenal glands to produce cortisol.

Agnosia (83) – The inability, in the absence of sensory problems, to recognize sounds, objects, people, etc.

Agonist (9) – A substance that binds to a receptor, activates it, and mimics the effects of a natural neurotransmitter.

Agraphia (84) – The loss of or difficulty with ability to write usually due to an acquired injury or neurological condition.

Alzheimer's disease (69) – A degenerative condition that results in a progressive loss of memory and ability to take care of oneself and is the most common form of dementia.

Amygdala (48) – A portion of the temporal lobe involved in processing emotion such as fear, anxiety, and aggression and to help form memories.

Amyloid plaques (69) - Abnormal clumps of protein that are thought to be one of the hallmark features of Alzheimer's disease.

Anatomy (7) – The way something looks such as the way brain structures look.

Angular gyrus (52) – A portion of the parietal lobe that is involved in reading, processing written information, mathematical operations, copying, drawing, and more.

Antagonist (9) – A substance that binds to a receptor and blocks or dampens the effects of a natural neurotransmitter.

Aphasia (79) - A disorder of language that impacts the ability to communicate.

Apraxia (80) - The inability to perform a task or movement despite understanding what is being asked and having the physical ability to perform the task.

Arachnoid mater (30) – The middle layer of the meninges that provides cushioning but is not as tough as the dura mater.

Arcuate fasciculus (27) – A bundle of nerve fibers that connect Broca's area and Wernicke's area to allow for the passing of information between them.

Ascending tract (21) – A nerve pathway that carries sensory information from the body to the brain.

Astrocyte (13) – A type of glial cell in the central nervous system that helps to manage the availability of essential nutrients (such as glucose) for neurons to make sure the neuron, including nucleus, has energy and can function properly.

Ataxia (73) – A neurological symptom characterized by lack of voluntary movement.

Atrophy (69) - Shrinking of organs, tissue, or muscle.

Attention (5) - Cognitive processes that generally refer to the ability to receive and process incoming information.

Auditory cortex (48) – A portion of the temporal lobe that processes auditory information including pitch and direction.

Autoimmune disorder (71) - A condition in which the body's immune system attacks healthy parts of the body.

Autonomic nervous system (3) – Part of the peripheral nervous system that helps to control involuntary bodily functions such as digestion.

Axial view (67) - An anatomical view that divides the body into upper and lower parts.

Axon (10) – The portion of a neuron that conducts electrical impulses.

Axon terminal (10) – The end of an axon where neurotransmitters are then released.

Basal ganglia (40) – A group of brain structures, including the caudate nucleus, substantia nigra, globus pallidus, nucleus accumbens, putamen, and subthalamic nucleus, that are involved in movement, emotion, and cognition.

Bipolar disorder (81) - A mental health diagnosis that includes fluctuations between periods of major depression and mania or hypomania (which are elevations in mood) and is often treated with medication and psychotherapy.

Brain (2) - An organ that contains nerve tissue and helps to control various body processes and functions.

Brainstem (23) – The lower part of the brain that connects the cortex with the spinal cord and helps to control essential survival functions.

Broca's aphasia (79) – A type of aphasia in which Broca's area is damaged causing difficulty with language production.

Broca's area (42) – A portion of the (typically left) frontal lobe that is associated with speech and language production.

Capgrass syndrome (78) – A disorder in which a person believes that someone close to them has been replaced by an identical imposter.

Cell body (9) – Also known as the soma, this part of the neuron contains the nucleus, integrates incoming signals from dendrites, and generates outgoing signals to the axon.

Central nervous system (2) – The part of the nervous system that includes the brain and spinal cord.

Cerebellum (36) – A brain structure located at the back of the brain and under the occipital lobe that is involved in voluntary movement and nondeclarative learning and memory.

Cerebral cortex (17) – The outer layer of the brain which is composed of grey matter and is responsible for higher order functions.

Cerebrospinal fluid (17) – A clear bodily fluid found in the brain and spinal cord that aids in cushioning and protection, removes waste, and provides nutrients.

Cerebrovascular system (30) – A network of blood vessels that supplies blood to the brain via capillaries, veins, and arteries.

Chorea (77) – Hyperkinetic movement involving rapid, sudden, and unpredictable movements.

Choroid plexus (29) – An area in the brain that produce cerebrospinal fluid.

Cingulate cortex (39) – A brain structure involved in emotional formation and processing, pain processing, and higher order cognition.

Cognition (5) – The mental processes related to thinking, knowing, remembering, etc.

Conduction aphasia (85) - A type of aphasia in which repetition is typically impaired while fluency and reception remains intact and is often caused by damage to the arcuate fasciculus.

Contralaterality (46) – The concept that one side of the brain controls or effects the opposite side of the body.

Coronal view (69) – An anatomical view that divides the body into front and back parts.

Corpus callosum (27) – A thick bundle of nerve fibers connecting both hemispheres to allow for the passing of information between them.

Cortical blindness (83) – Vision loss, despite having heathy eyes and optic nerves, due to damage to the primary visual cortex.

Corticospinal tract (22) – A descending pathway that carries motor information, such as voluntary motor control information, to the spinal cord.

Corticotropin-releasing hormone (61) – A hormone that is produced by the hypothalamus that plays a role in the stress response by stimulating the release of ACTH by the pituitary gland.

Cortisol (60)– A hormone that helps to regulate metabolism, immunes response, is involved in the body's stress response, and is also known as "the stress hormone."

Cranial nerves (31) – The twelve pairs of nerves that emerge from the brain and are involved in numerous automatic sensory and motor functions.

Cushing's syndrome (75) – A disorder characterized by excessive and prolonged exposure to cortisol with symptoms ranging from weight gain to high blood pressure to thinning skin.

Death (73) - To cease living.

Degenerative condition (67) – A disease or disorder characterized by deterioration over time.

Dementia (69) - A broad diagnostic term to describe a condition that involves a decline in cognition that interferes with ability to take care of oneself and includes specific conditions such as Alzheimer's disease, vascular dementia, frontotemporal dementia, and dementia with Lewy bodies.

Dendrite (9) – The branches of a neuron that receive signals from other neurons which are then transmitted towards the cell body.

Dentate gyrus (17) – A portion of the hippocampus involved in forming new episodic memories and one of the few areas of the brain in which neurogenesis occurs.

Descending tract (21) – A nerve pathway that carries motor information from the brain to the muscles of the body.

Diabetes (75) – A chronic medical condition in which the body either does not produce insulin or does not use insulin properly.

Diffuse (78) – Broad or widespread.

Disease modifying treatment (71) - A treatment that attempts to slow the progression or reverse a disease process by treating the underlying cause.

Dominant hemisphere (20) – The hemisphere of the brain that is more involved in controlling certain, but not all, functions such as language.

Dopamine (8) – A neurotransmitter involved in many functions including movement, reward, motivation, learning, and pleasure.

Dorsal column-medial lemniscus pathway (22) – An ascending pathway in the spinal cord that sends sensory information, such as fine touch and proprioception, from the body to the brain.

Dorsal stream (51) – A pathway in the brain from the occipital lobe to the parietal lobe involved in the processing of spatial information such as the movement of something and is also known as the "where pathway."

Dorsolateral prefrontal cortex (41) – A portion of the prefrontal cortex involved in higher order cognition such as problem solving, judgment, decision making, and memory.

Dura mater (30) – The outer layer of the meninges that is the toughest and provides protection to the brain and spinal cord.

Dyscalculia (84) – A learning disorder of math such as understanding number-based concepts, performing calculations, and understanding math facts and formulae.

Dysgraphia (84) – A learning disorder of writing such as with spelling, putting thoughts on paper, or poor handwriting.

Endocrine system (59) – A network of glands and organs involved in the production, storage, and release of hormones.

Ependymal cell (16) – A type of glial cell lining the ventricles of the brain and spinal cord that produce and help to circulate cerebrospinal fluid.

Etiology (64) - The presumed cause.

Executive functioning (6) - Cognitive processes that generally refers to higher order cognition including our ability to plan, make judgements, reason, make decisions, solve problems, etc.

Fight or flight response (3) – The bodily reaction that occurs in response to a threat or perceived harmful event which prepares the body to either confront or flee from the threat.

Finger agnosia (84) – The inability to distinguish individual fingers on one's own hand or the hands of others and is often seen in Gerstmann syndrome and associated parietal lobe damage and conditions.

Frontal eye fields (43) – A portion of the frontal lobe helping to control voluntary eye movements.

Frontal lobe (41) – The part of the cerebral cortex located at the front of the brain associated with higher order cognition, emotion, personality, and more.

Frontotemporal dementia (70) – A group of disorders caused by a degeneration of cells in the frontal and temporal lobes.

Function (7) - The way something works such as how brain structures work.

Gamma-aminobutyric acid (8) – Also known as GABA, it is the primary inhibitory neurotransmitter in the central nervous system involved in regulating stress, anxiety, and sleep.

Generalized anxiety disorder (81) - A mental health diagnosis characterized by excessive worry about a variety of topics which is difficult to control and is often treated with medication and/or psychotherapy.

Gerstmann syndrome (84) – A neurological condition due to damage to the left angular gyrus that often results in a constellation of symptoms including left-right disorientation, finger agnosia, a/dysgraphia, and a/dyscalculia.

Gland (59) – An organ or tissue that produces and releases substances such as enzymes or hormones.

Glial cell (13) – A type of cell in the nervous system that helps to protect and support neurons.

Glutamate (8) – The primary excitatory neurotransmitter in the central nervous system involved in learning, memory, and other cognitive functions.

Granule cell (18) – A type of small cell found in various areas of the brain such as the dentate gyrus.

Grey matter (12) – Portions of the central nervous system that contain mostly neuronal cell bodies, dendrites, and unmyelinated axons.

Hemi-neglect (74) – A lack of awareness of one side of space or body.

Hemorrhagic stroke (66) – A type of stroke caused by the rupturing of a blood vessel causing bleeding in the brain.

Hippocampus (47) – A region of the brain located in the temporal lobe that is involved in spatial navigation and helps to form and store memories.

Homeostasis (14) – A process by which a cell or organism attempts to maintain a stable internal environment or "balance."

Homunculus (45) – A visual, mapping representation of the areas of the cortex dedicated to processing motor or sensory information for different parts of the body whereby body areas that require more information are drawn disproportionality larger.

Hormone (59) – A chemical messenger produced by glands in the endocrine system that aid in regulating different processes in the body.

Hypothalamic-Pituitary-Adrenal Axis (HPA Axis; 60) – A portion of the stress response system that functions as a more delayed response to a threat or challenge and involves the release of cortisol.

Huntington's disease (77) – A neurodegenerative condition with a highly genetic contribution characterized by hyperkinetic movement.

Hydrocephalus (65) – A condition in which cerebrospinal fluid accumulates causing increased pressure on the brain.

Hyperkinetic (77) – Symptoms or behaviors characterized by excessive and involuntary movements commonly seen in Huntington's disease.

Hypertension (75) – A medical condition in which arterial blood pressure is elevated.

Hypokinetic (77) – Symptoms or behaviors characterized by reduced movement and commonly seen in Parkinson's disease.

Hypothalamus (37) – A brain structure located under the thalamus that helps to regulate many functions related to homeostasis such as hunger, sleep, and thirst.

Hypothyroidism (76) – A condition in which the thyroid gland does not produce enough hormone and can lead to fatigue, weight gain, depression, and more.

Inferior colliculus (35) – A portion of the tectum that helps to automatically orient to auditory information.

Insular cortex (54) – A region in the cerebral cortex that is found deep near the juncture of the frontal, parietal, and temporal lobes and is involved in self-awareness, social-emotional processing, homeostasis, motor control, and visceral reactions.

Ischemic stroke (66) – A type of stroke caused by a blockage in an artery that supplies blood to the brain.

Kluver-Bucy syndrome (81) – A rare condition in which the bilateral anterior temporal lobes (including amygdalae) are damaged resulting in symptoms such as hyperorality, reduced fear, reduced aggression, and hypersexuality.

Language functioning (6) - Cognitive processes that generally refers to the ability to express and receive communication.

Left hemisphere (20) – The left half of the cerebral cortex.

Left-right disorientation (84) – A condition in which a person cannot distinguish between their left and right sides and is often seen in Gerstmann syndrome and associated parietal lobe damage and conditions.

Limbic system (57) – A network of structures in brain involved in regulating emotions, behavior, and memory.

Lobe (25) – A division of the cerebral cortex into distinct areas.

Locked-in syndrome (73) – A neurological condition in which a person in conscious and aware but cannot move or communicate due to paralysis of voluntary muscles aside from the eyes.

Major depressive disorder (81) - A mental health diagnosis that is characterized by intense and persistent feelings of sadness and associated symptoms such as loss of interest, appetite/sleep changes, and more which is often treated by medication and/or psychotherapy.

Medulla (33) – A brainstem structure that connects to the spinal cords and helps to regulate many basic involuntary and life functions such as breathing and heart rate.

Memory (6) - Cognitive processes that generally refer to the ability to input, store, and retrieve information.

Meninges (29) – Protective membranes that surround the brain and spinal cord.

Medial prefrontal cortex (41) – A portion of the prefrontal cortex involved in motivation, emotion, judgment, and motor function.

Mesocortical system (61)– A pathway in brain that utilizes dopamine and is involved in motivation, decision-making, and learning.

Mesolimbic system (61) – A pathway in brain that utilizes dopamine and is involved in reward, motivation, and learning.

Microglia (16) – A type of glial cell in the central nervous system that is involved in the immune response.

Midbrain (35) – A brainstem structure above the pons that is involved in movement, reward, and pain interpretation.

Multiple sclerosis (71) – An autoimmune disease that affects the central nervous system in which the immune system attacks the myelin sheath.

Multipolar neuron (7) – A type of neuron with one axon and numerous dendrites.

Myelin sheath (10) – The fatty insulation of an axon that helps to speed up electrical impulses.

Nervous system (2) – The complex network of cells and nerves that transmit signals to different areas in the body.

Neurofibrillary tangles (69) - Abnormal accumulations of a protein known as tau which are a characteristic feature of neurodegenerative disorders such as Alzheimer's disease.

Neurogenesis (4) – The process in which new neurons are created.

Neuron (7) – A cell that is the basic building block of the nervous system.

Neurotransmitter (8) – A chemical messenger in the nervous system.

Nodes of Ranvier (10) – Gaps in the myelin sheath allowing for faster transmission of electrical impulses.

Nondeclarative memory (37) – A type of memory that does not require conscious thought.

Norepinephrine (8) – A neurotransmitter that is involved in stress response, attention, and arousal.

Nucleus (9) – The portion of a neuronal cell body that contains genetic material.

Occipital lobe (50) – The part of the cerebral cortex located at the back of the brain associated with processing visual information.

Olfactory bulb (46) – A structure found underneath the frontal lobes that helps to transmit information about olfaction (smell).

Oligodendrocyte (15) – A type of glial cell in the central nervous system that produces the myelin sheath.

Orbitofrontal prefrontal cortex (41) – A portion of the prefrontal cortex involved in organization, inhibition, and emotion regulation.

Oxytocin (8) – A neurotransmitter also known as the "love hormone" as it is involved in social bonding, regulating emotions, empathy, and parental behaviors.

Paramnesia (78) – A condition which memories are in some way distorted such that a person may have trouble distinguishing between reality and fantasy.

Parasympathetic nervous system (3) – Part of the autonomic nervous system that helps the body relax and return to homeostasis.

Parietal lobe (51) - The part of the cerebral cortex located at the top and back of the brain associated with processing sensory information.

Parkinson's disease (77) – A neurodegenerative condition characterized by hypokinetic movement.

Peripheral nervous system (2) – The part of the nervous system outside of the brain and spinal cord which connects the central nervous system to the organs and limbs.

Pia mater (30) – The thinnest and innermost layer of the meninges that directly covers the brain and spinal cord.

Pituitary adenoma (75) – A tumor (typically benign) that comes from the pituitary gland and can affect hormone production.

Pituitary gland (38) – A small gland located near the hypothalamus that produces and releases numerous hormones and helps to regulate other glands.

Pons (34) – A brainstem structure between the medulla and midbrain that is involved in basic survival functions such as breathing and arousal.

Posttrauamtic stress disorder (81) - A mental health diagnosis stemming from the experience of a traumatic event that may include symptoms such as reexperiencing, intrusive thoughts, avoidance of aspects of the trauma, hypervigilance, and is often treated with psychotherapy and sometimes medication.

Post-synaptic neuron (10) – The neuron receiving the signal across the synapse.

Prader-Willi syndrome (75) – A genetic condition caused by loss of function to specific genes on chromosome 15 and is characterized by a constant feeling of hunger.

Prefrontal cortex (41) – A portion of the frontal lobes that is involved in higher order cognition such as problem solving, decision making, impulse control, as well as, social behavior, emotion, and personality.

Premotor cortex (44) – A portion of the frontal lobe helping to coordinate and sequence movements.

Pre-synaptic neuron (10) – The neuron sending the signal across the synapse.

Primary motor cortex (45) – A portion of the frontal lobe that is involved in voluntary movement control.

Primary visual cortex (50) – A portion of the occipital lobe that is involved in the processing of visual information such as shape, color, motion, and depth perception.

Procedural memory (37) – A type of nondeclarative memory that involves the learning and memory of procedures such as riding a bicycle, driving a car, typing, or playing an instrument.

Processing speed (6) - A cognitive process that generally refers to the ability to use that information quickly and efficiently.

Proprioception (22) – The sense of the position of one's body in space.

Reduplicative paramnesia (78) – A rare delusion in which a person believes that a person or place has been duplicated and exists in multiple locations simultaneously.

Refractory period (10) – A period of time following an action potential in which a neuron cannot fire again.

Reticular formation (56) – A network of neurons found throughout the brainstem that helps to maintain alertness, consciousness, regulation of sleep-wake cycles, and more.

Reuptake (11) – A process in which neurotransmitters that have been released into the synapse are then reabsorbed by the pre-synaptic neuron.

Right hemisphere (20) – The right half of the cerebral cortex.

Rubrospinal tract (22) – A descending pathway that carries motor information, such as coordination of movement, to the spinal cord.

Sagittal (26) – An anatomical view that divides the body into right and left parts.

Schizophrenia (79) - A mental health diagnosis that is characterized by psychotic symptoms such as hallucinations and delusions, changes in mood and affect, and cognitive difficulties and is often treated with use of medication and sometimes psychotherapy.

Schwann cell (16) – A type of glial cell in the peripheral nervous system that produces the myelin sheath.

Seizure (84) - A sudden, uncontrollable burst or disruption of electrical activity in the brain.

Sensorimotor functioning (5) - Cognitive processes that generally describes the ability to use sensory information and integrate it into motor function.

Serotonin (8) – A neurotransmitter involved in many functions including mood, sleep, and appetite.

Skull (29) – The bony structure that encases and protects the brain.

Somatic nervous system (3) – Part of the peripheral nervous system that helps to controls voluntary movement.

Somatosensory cortex (51) – A portion of the parietal lobe that helps to process touch, pain, temperature, and more.

Spinal cord (2) – The long, thin structure composed of tissue that extends from the brainstem down the vertebral column and helps to transmit nerve signals between the brain and the rest of the body.

Spinal cord injury (71) – Damage to the spinal cord that results in loss of function.

Spinocerebellar tract (22) – An ascending pathway in the spinal cord that carries sensory information, such as proprioception, from the body to the brain.

Spinothalamic tract (22) – An ascending pathway in the spinal cord that carries sensory information, such as pain and temperature, to the brain.

Split-brain (84) - The inability for the left and right hemispheres to pass information to each other often due to a surgery and damage to the corpus callosum.

Stroke (66) – A condition in which the blood supply of the brain is interrupted or reduced which deprives the brain tissue of oxygen and nutrients.

Subarachnoid hemorrhage (31) – Bleeding into the subarachnoid space, for example from a ruptured aneurysm.

Substantia nigra (36) – A part of the tegmentum that houses cells that produce dopamine and is implicated in Parkinson's disease.

Superior colliculus (35) – A portion of the tectum that helps to automatically orient to visual information.

Supramarginal gyrus (53) – A portion of the parietal lobe involved in the processing of written information such as assigning meaning to words.

Sympathetic nervous system (3) – Part of the autonomic nervous system that helps the body mobilize resources and to respond to a threat or challenge by increasing heart rate, inhibiting digestion, etc.

Sympatho-Adreno-Medullary axis (SAM axis; 59) – A portion of the stress response system that is typically one of the first, or initial, responses to a threat or challenge.

Synapse (10) – The space between two neurons where neurotransmitters are released to transmit information.

Tectum (35) – A portion of the midbrain involved in visual and auditory reflexes.

Tegmentum (35) – A portion of the midbrain involved in movement and reward.

Temporal lobe (47) – A region of the cerebral cortex located under the lateral fissure in both hemispheres of the brain.

Thalamic stroke (74) – A stroke occurring in the thalamus which can result in impairments sensory and motor information transmission.

Thalamus (37) – A brain structure that serves as a relay station for sensory and motor information to other areas of the brain.

Tract (21) – A bundle of nerve fibers in the central nervous system that transmits signals.

Trauma (67) – Physical injury or damage to the body such has a traumatic brain injury.

Tumor (67) – An abnormal mass of tissue that can be either cancerous or non-cancerous.

Vascular dementia (67) - A type of dementia often due to significant cardiovascular events such as stroke or heart attack and/or risk factors such as hypertension, diabetes, sleep apnea, and more.

Ventral stream (51) – A pathway in the brain from the occipital lobe to the temporal lobe involved in object recognition and is also known as the "what pathway."

Ventral tegmental area (35) – A portion of the tegmentum that is involved in reward and the release of dopamine.

Ventricle (29) – A cavity in the brain that contains cerebrospinal fluid.

Vesicle (10) - A structure in which neurotransmitters are stored and then released.

Vestibulospinal tract (22) – A descending pathway that carries motor information, such as balance and posture, to the spinal cord.

Visuospatial functioning (6) - Cognitive processes that generally refers to the ability to identify and process visual information and spatial relationships.

Wernicke's aphasia (82) – A type of aphasia in which Wernicke's area is damaged causing difficulty with language reception.

Wernicke's area (49) – A portion of the (typically left) temporal lobe that is associated with speech and language reception.

White matter (12) – Portions of the central nervous system that contain mostly myelinated axons.

AUTHOR'S NOTE AND REFERENCES

My goal for those reading was to feel immersed in the company that is NeuroVoy Industries. To capture this and to keep the reader engaged and text flowing, I opted to comment on references used here in this portion of the book rather than use traditional in-text citing of work that I typically would use in an academic paper, for example. Therefore, in addition to my own personal and clinical experiences, I would like to indicate that information was used from the following sources in aiding my creation of this book, developing the glossary, and I am grateful to the authors of those sources for having a profound impact on my clinical and academic career. I would absolutely recommend any reader to review these texts for more information!

- **Clinical Neuropsychology: A Pocket Handbook for Assessment** (2014) by Dr. Michael W. Parsons & Dr. Thomas A. Hammeke. This text was used primarily in the creation of sections three through six of the handbook.

- **Exploring Psychology** (2019) by Dr. David G. Myers & Dr. Nathan DeWall. This text was used primarily in the creation of sections one and two of the handbook.

- **Health Psychology** (2023) by Dr. Richard O. Straub. This text was used primarily in the creation of section five of the handbook.

- **Neuroanatomy Text and Atlas** (2012) by Dr. John H. Martin was used primarily in the creation of sections one through three of the handbook.

- **The Clinical Neuropsychology Study Guide and Board Review** (2020) by Dr. Kirk J. Stucky, Dr. Michael W. Kirkwood, Dr. Jacobus Donders, & Dr. Christine Liff. This text was used primarily in the creation of sections two through six of the handbook.

- **The Little Black Book of Neuropsychology** (2016) by Dr. Michael R. Schoenberg & Dr. James G. Scott. This text was used primarily used in the creation of sections three through six of the handbook.

Lastly, Brandcrowd.com was used to create the NeuroVoy Industries company logo and Shutterstock.com was used for all other images found in this handbook. These images are all being used under their respective licensing terms. A huge thank you to those two websites for the easy-to-use platforms.

ACKNOWLEDGMENTS

To my son, Colin, I will love you more than you will ever know. You've made me into a better person and I thank you for that. I am so excited to see you grow and experience the world. I will always be with you, even when I'm not. To my wife, Christine, thank you for supporting my professional and personal goals. You have, time and time again, sacrificed in order for me to achieve my dreams. Thank you. You are the best friend, wife, and mother to Colin that I could ask for. Most of us have some comic book hero, movie star, or athlete that we look up to as children and that is certainly the case for me. But, my true superheroes and the people I look up to more than anyone are my parents, Connie and Jerry. Thank you for the unwavering, continued, unconditional love and support. You provided everything we needed and wanted as children (and now) and your sacrifice and hard work is something I will pass along to Colin. To my siblings, Cara and Brian, thank you for always being there for your big brother. We laugh and joke in a way that is unique to our bond and is one that can never be replaced. I cherish every moment we have been able to share. I would like to also thank my extended family including my grandparents, aunts, uncles, and cousins from the McKay and Spencer side (and beyond), as well as, my newer family through my wife including the Nelson and Harmon families. Further, I have had the honor of forming so many positive relationships and friendships in my life and I cherish each and every one of you, both local and afar, older and newer. I would especially like to thank my best friends, Matt Swanseger and Drew Lam, and their families for treating me as one of their own. Matt, we've been best friends since kindergarten which I think is a relatively rare feat these days. You always find a way to make me laugh and I love reminiscing about our escapades over a cold one. Drew, we didn't meet until graduate school, but I'm so happy we did. I know I can always count on you for an open ear. I always look forward to breaking down the latest TV show and making at least one trade during fantasy season.

From an educational and career perspective, so many have shaped who I am as an academic and as a clinician today. To Mr. Russ Taylor, my high school psychology teacher, thank you for laying the foundation for my interest in psychology. The way you taught comes out in the way I teach today. Thank you to Dr. Eric Corty, who took me under his wing at Penn State Behrend, taught me how to conduct research, and prepared me for graduate school. To Drs. David Tokar, Kevin Kaut, Gino Peluso, Eric Hayden, and Chris Buzzelli, thank you for supporting and teaching me during a time of transition in my life, when I decided to become a neuropsychologist in a program that was not built for training one. You helped to build my path. A special thank you to Dr. Kevin Kaut, who reviewed this book and provided extremely helpful and thoughtful feedback prior to publication. To my cohorts, staff, mentors, colleagues, and friends of my training and employment following graduate school, including from the Dallas VA, Salem VA, Meadville Medical Center, Erie VA, and Mercyhurst University, thank you for taking a chance on me, teaching me, supporting me, forging friendships, and helping to shape who I am today. To my students, I hope that you will always take something positive and informative from my classes. Thank you for allowing me to do what I love to do, even if it includes the occasional terrible joke (OK, maybe more than occasional). Finally, to my patients, thank you for the privilege of getting to be your doctor, part of your care, and helping as best I can. Again, thank you all. I appreciate your continued support more than you know.

ABOUT THE AUTHOR

Dr. Derek McKay is a licensed clinical psychologist and is board-certified in neuropsychology. He earned his B.S. in Psychology from Penn State Behrend and his M.A. and Ph.D. in Counseling Psychology from the University of Akron. He completed a one-year APA-accredited internship in the clinical neuropsychology track at the VA North Texas Health Care System in Dallas, Texas. He then completed a two-year APA-accredited postdoctoral residency program in clinical neuropsychology at the Salem VA Medical Center in Salem, Virginia. Currently, Dr. McKay works full-time as an assistant professor of psychology at Mercyhurst University in Erie, Pennsylvania, and teaches courses that are related to clinical work and neuropsychology. He enjoys teaching students about the neuroanatomy and the brain, brain health, mental health and medical conditions, and the complex connection between them. Additionally, Dr. McKay owns and operates Flagship Neuropsychology PLLC where he conducts neuropsychological evaluations on a part-time basis. Dr. McKay also works as a fee-basis neuropsychologist for the Erie VA Medical Center. Dr. McKay's research and clinical interests are neurological disorders, autoimmune disorders, and dementia. He is also interested in models of personality, career interests, and subjective well-being. Dr. McKay enjoys spending time with his son, wife, family, and friends. He also enjoys trying new food and beer, traveling, enjoying the outdoors, and watching football.

FOR MORE INFORMATION

Please visit www.flagshipneuropsychology.com for more information and NeuroVoy Industries products/merchandise! You will also find more information regarding Dr. McKay's clinical practice, academic work, philanthropy, and more. Thank you all for the support!

Printed in the USA
CPSIA information can be obtained
at www.ICGtesting.com
LVHW052158240724
786350LV00019B/485